Isolation

– solitude, secrets, atoms . . .

Edited by: *David C. Gershlick*
              Janet Gibson
              Harshad K. D. H. Bhadeshia

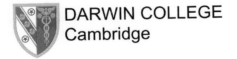

DARWIN COLLEGE
Cambridge

ISOLATION – solitude, secrets, atoms . . .

Arising from the 2023 Darwin College Lectures, this book presents eight essays on the theme of "isolation", each written by an academic with an exceptional passion for the creation and dissemination of knowledge. Together, the essays constitute an interdisciplinary volume spanning the humanities and sciences, designed for a wide appeal.

Isolation features contributions by Christine van Ruymbeke, Amy Nethery, Adrian Kent, Jane Francis, Philip Jones, Arik Kershenbaum, Heonik Kwon and Amrita Narlikar, all distinguished academics and excellent communicators.

The subjects covered range from solitude, asylum, secrets, a continent, light, aliens, a country, and trade, but by intent, not *the* pandemic.

PUBLISHED BY DARWIN COLLEGE
Silver Street, Cambridge CB3 9EU United Kingdom

https://www.darwin.cam.ac.uk

© Darwin College, Cambridge, 2023

First published 2023

Isolation – solitude, secrets, atoms . . .
        edited by David Gershlick, Janet Gibson and Harshad Bhadeshia
Includes index

ISBN 9798353788713 Paperback

Typeset in Computer Modern font
by the editors

Contents

Preface

We think this book will interest you – all of the articles are written by academics who impress not just by their depth of knowledge, but more than that, each author is a communicator *par excellence*, with an innate ability to inspire across the disciplines.

Many trains of thought can be triggered by the word *isolation* some of which can delight, for example, the idea of being interviewed by the BBC on the programme *Desert Island Discs* and then agreeing to be abandoned on a deserted island with little more than a couple of possessions as the reward. Indeed, the word derives from the Latin *insula* meaning 'island'. Being shipwrecked on a similar location does not, however, appeal.

One form of solitude, when a person chooses to separate themselves from society and locate in a remote, far-away place, effectively living in a self-imposed silent order while contemplating the meaning of life, still features in modern times. But there is a secret that spans centuries, with content that is mystifying: the solitary student, the Turtle's wife and much more. This is the forgotten Persian literature including poetry, which was nurtured, destroyed and reborn. Elementary emotions are addressed, sometimes in prose – can affection remain intact and untwisted only if it is divine? Christine van Ruymbeke begins with a Mirror for Princes, containing fables that advise busy would-be-rulers. The ruler is condemned to loneliness through an inability to trust. But at the same time, a good companion is said to be better than solitude. Are these contradictions to be reconciled? We are then treated to an enchanting 5th-century story (poem) told to a king by the daughter of the Slav king. The poem has meaning, which until Christine, had not been decoded. It is about the only daughter of an aged king; she sets clever and sometimes terrible barriers to suitors, which must be overcome to win her, but where the cost of failure is painful. Eventually, a successful candidate emerges, but the silent exchanges during the trials leave us (and indeed, the mythical father) at a loss to understand why the suitor won her heart. Christine solves this mystery in a manner that captivates and charms: a smile-inducing exposition in the best traditions of wonder.

The 1717 Transportation Act enacted during the reign of George I of Great Britain, had the purpose of compensating for the ineffectiveness of contemporary law to deal with evil-disposed people. Instead, it allowed felons and paupers to be transported to any part of North America for seven years. The Act specifically excluded application to that part of Great Britain called Scotland. Transportation to America came to a halt after the war of independence. Experiments were therefore carried out to put prisoners into ship hulks along the coastline, but this led to appalling conditions on board and the death of many through disease. In 1788, the British Government created a penal colony in Australia, leading eventually to the transportation of more than 160,000 convicts.

Modern Australia is perhaps the only democracy where neighbouring islands such as Manus and Nauru are used as a part of the government's zero-tolerance policy of deterring maritime asylum-seekers. Amy Nethery, through her research, explains the human consequences of this policy on Australia and Asia in general. Her essay makes poignant reading in the context of a variety of ideas considered by the current British Government to send immigrants who arrive illegally across the English Channel, to Ascension Island and St Helena, or transporting them to Rwanda. Indeed, this kind of thinking is not unique to the current government – Tony Blair, the Labour Prime Minister and his colleague David Blunkett proposed the setting up of special centres outside Europe, near conflict areas, so that refugees could return to their countries once the conflict ended. This was in 2003, coinciding with the invasion of Iraq which was not supported by the United Nations.

The secrecy imposed on detention centres in Australia has had unimaginable consequences. But the concept of secrecy has particular significance in this age of surreptitious data collection and rapid communications. Of course, the speed of information transfer is limited by that of light – a space craft on Mars has its transmission completely secret to all on Earth for many minutes, until the radio signal reaches there. So any transmitted information is *temporarily* secret in all regions of space where the signal has yet to reach. Adrian Kent in his essay attempts to explain how quantum information differs from the classical form; so, for example, it becomes possible in the quantum world to make unforgeable bank-notes. Similarly, the idea that a quantum state can be measured precisely is lost, which makes cloning impossible. Even if a completely secure quantum system of information-guarding is developed and implemented, could new laws of physics be discovered that would render it transparent? Some of the ideas that Adrian writes about are difficult to appreciate for those trained to think classical, but the effort is there to make the subject digestible for those who persevere. Would it not be fun to be able to hold a conversation with Adrian's Alice and Bob, and the evil Eve, in a scenario where each is granted immunity in exchange for revelations, whether they be about coin flipping, the fabrication of video evidence or attempts to influence the local gravity field by dragging in a neutron star?

That Antarctica is often forgotten as a continent is not surprising – there are no humans that make their homes there and what lies under and around the ice-cover is probably more fascinating to ordinary mortals than the continent itself. Even with modern paraphernalia, it is a difficult place to visit, and international treaties cover the presence of humans on the continent. On the other hand, children will be much more aware of its existence because it features regularly in cartoons and has the largest population of penguins in the world, comprising some twenty million breeding pairs.

It was about 40 years ago that Jonathan Shanklin of the British Antarctic Survey in Cambridge, made his monumental discovery of the ozone hole, a discovery that changed global perceptions and led eventually to its healing. His analysis used scientific instruments based in Antarctica; there are around fifty, occupied scientific bases there.

Jane Francis rightly points out that what we now call the Antarctica, was a part of a supercontinent, Gondwana, during the Jurassic period. Of all the planets in the Solar System, Earth is the only one where the solid surface is not continuous but consists of "loosely" connected tectonic plates that can migrate independently over geological time. This migration caused the fracture of Gondwana; Antarctica is just a detached bit. It is now hard to imagine that it was once contiguous with forested land masses, vestiges of which exist today. Jane explains why the continent is now covered with ice, although unlikely as it seems, that may change. A person drinks about 50 tonnes of water during their lifetime; ice weighing 150 billion tonnes is melting annually from Antarctica.

It might surprise that light exerts a force – after all, the speed of light is fixed so it cannot accelerate. Moving photons do, however, carry energy and hence have a momentum. The Viking spacecraft would have missed its Mars orbit by many thousands of kilometres if the force of light from the Sun, on its trajectory, had not been taken into account. But the force of light is small, so we as large objects on Earth, do not feel it. Very small objects do, but the minuteness of the force meant that its consequences were at first neglected (or dismissed). Of all places, it was at the Bell Telephone Laboratories that Arthur Ashkin calculated that a laser could be used to concentrate light and stimulate an observable "quite large" effect. And as explained so elegantly by Philip Jones, that is where the story of what we now call "optical tweezers" begins. It becomes possible, for example, to measure the elastic properties of DNA and there are many more biological applications envisioned. Philip almost ends with a salutary quotation by Yogi Berra (really) about pontificating the future, but finishes with a safe estimate of the future.

Our own existence is sometimes justified by assuming that humans are exceptional, but the frame of reference is limited to the handful of planets that whirl around our Sun. What would it mean to discover an alien? Not five hands or eyes at the back of the head, but rather, as Arik Kershanbaum puts it, some entity with whom we can share our deepest thoughts. The question of whether life exists out there is not trivial once we account for our need to connect with others. The discovery of a worm on the planet Arrakis would therefore not be inspiring. In the translated words of Wittgenstein, "if a lion could speak, we could not understand him" – the nature of language requires shared experiences or feelings. Arik develops the concept of the human response to extraplanetary life with a clarity that goes far beyond what fictional imagination can conjure. His notion of the existence of alien life is not based on the usual conjectures about the inevitability of something happening somewhere by chance, given the vast numbers of planetary bodies that may exist out there.

A certain country has chosen to be, or has fallen into the role of being, the sole guardian of socialism as a balance to uncontrolled American imperialism following the end of the Cold War and the Chinese focus on economic growth rather than ideals. North Korea is the only country that has survived in its self-identified role by a military-first philosophy and unique also in having a hereditary leadership in spite of the role. Some in the West believe that North Korea is the most dangerous nation in the world, perhaps because it is nuclear-armed, without remembering that it once was threatened with an overwhelming nuclear bombardment by MacArthur. The mechanism by which North Korea in some ways adopted solitude, is unravelled with great care by Heonik Kwon, emphasising the connections that led eventually to its current state. How much do we believe that the people there see the country as a paradise on Earth? And does that belief matter?

There were many devastating consequences of the decade-long Great Depression (1929-1939), but an interesting feature was that those countries that did not engage in international trade, suffered little. The Russian economy at the time grew through heavy industrialisation, so much so that people emigrated to Russia from Finland and Germany. The isolationist attempt to protect economies after the Depression exacerbated the effects. After the Depression, there was a push to boost international trade as a means of increasing national wealth. There is good evidence, both from those times and indeed, from the modern development of the Chinese economy, that vast numbers of people emerged from abject poverty due to participation in international trade. However, trade issues never have been isolated from political aspirations, as illustrated vividly by Brexit in the U.K. and the arbitrary U.S. trade tariffs imposed during its recent political perturbations. Amrita Narlikar presents a picture of how the perceptions of open trade might have to be tempered by focussing on "like-minded" economies. After all, trade can so easily be used to inflict pain. The story is necessarily intwined with many strands of complexity, but she leaves us with an intriguing thought, that war can be avoided if a form of limited isolationism is permitted, in which countries would have a choice on the level and timing of engagement.

When the World Health Organisation on the 9th of January 2020 noted a mysterious outbreak in Wuhan, little did the world know of the coming catastrophes. Isolation soon became a term of dread in many different ways; medical staff needed shielding from the virus, badly affected patients could not be allowed to see relatives, and much more. We decided at the outset, that we would not cover the pandemic in the context of this series of lectures, because there exist some 600,000 academic articles published since the WHO announcement, not counting the media coverage.

There are topics that we would have liked to cover, but alas, there are constraints. To tantalise you, here is some suggested reading. Populations that do not grow much or migrate, may lack sophistication – the indigenous people of Tasmania developed just 24 types of tools (Rutherford 2022). What happens to babies who are raised in total isolation (Hood 2018)? Anil Seth (2021) has written a marvellous book on what it means to be a 'self' – it would be intriguing to know whether consciousness can be isolated, to establish that it is not just a figment of our fertile imagination.

The Darwin College Lecture Series, conceived originally by Andrew Fabian, is a gem that illustrates the advantages of non-curricular learning that does not respect boundaries. We, the editors and organisers of *Isolation*, are, to say the least, delighted to be given this opportunity to attract another cluster of fine academics to entertain our huge audiences with their knowledge, wit and more than anything else, a story to tell and retell.

We are particularly grateful to the Education and Research Committee of the College, chaired by Torsten Krude, for its efforts in shaping the Series, with the planning in each case beginning two years ahead of delivery. Much appreciated are: Espen Koht who leads with all of the technology associated with the live presentations; the University of Cambridge Audio-Visual services for recording the lectures; the numerous Fellows and Members of the College, adorned in their academic regalia, who help manage the enthusiastic audiences that attend these lectures; the staff of the Darwin Catering Team led by Ivan Higney; Laura Pellegrini who designed the lovely cover; Nigel Bowles who graciously volunteered to copy-edit the book; Michael Rands, the Master of the College for his support throughout, particularly when difficult decisions had to be made.

It goes without saying, that we thank all those who agreed without hesitation, to contribute both the lectures and the written masterpieces.

David C. Gershlick, Janet Gibson and Harshad K. D. H. Bhadeshia
Darwin College, Cambridge

Hood, B. (2018), *Development – mechanisms of change*, Cambridge University Press, eds T. Krude and S. T. Baker, Cambridge, U.K., chapter 4: Developing a sense of self, pp. 53–98.

Rutherford, A. (2022), *Enigmas*, Cambridge University Press, eds E. J. Ward and R. Reuvers, Cambridge, U.K., chapter 1: Human origins, pp. 10–39.

Seth, A. (2021), *Being you, a new science of consciousness*, Faber & Faber, London, U.K.

1 On escaping or not escaping solitude. Persian tales of turtles and pearls

CHRISTINE VAN RUYMBEKE
University of Cambridge

Abstract: Narratives speak volumes. As remarked by the philosopher Hannah Arendt, they are the only possible medium to express the complexity of philosophical or other conundrums. Often, the reader's effort to decode them, that is: the exercise itself, contains the pedagogy. This presentation examines two great Medieval Persian narrative works: the Book of Kalila and Dimna (کتاب کلیله و دمنه Ketab-e Kalile-o Demne) by Nasrollah Monshi and the Seven Portraits (هفت پیکر Haft Paykar) composed by Nezami Ganjavi. Should we escape or not escape solitude, that feeling experienced in a state of physical or mental isolation? Aren't friendship or love preferable, even at very high costs? My two authors-philosophers propose deadly serious situations . . . and leave us to work them out. Listen to what happens to the bored old King of the Monkeys up in his fig tree! Be baffled by the riddles of the Princess who sets such very high standards to her suitors! I will introduce you to my way of interacting with these two sophisticated narratives written for the highest social strata of the Persianate world; they address people whose expectations from literature were manifestly different from ours in this 21st century. Nevertheless, my aim is to see you leave this chapter with a huge smile on your face and a new way of looking at friendship and love, basking in a new awareness of what Persian tales do to us and how irresistible they are.

"To live alone one has to be a beast or god – says Aristotle.
But there's a third case: one has to be both – a *philosopher*
- F. Nietzsche, *Twilight of the Idols*, Maxim 3

Medieval and pre-modern Persian literature, especially its poetry, is one of the world literature's least studied and best kept secrets. Its quality and resonance place it on par with other great literary traditions and it rewards study beyond expectations. For a general awareness of a handful of great poets' names and verses, such as Mowlana Rumi, Hafez, Sa'di and naturally, Omar Khayyam, we are indebted to the pioneering efforts of great orientalists. These scholars are much criticised nowadays, nevertheless they are to be thanked for introducing these Persian classics to a Western audience. They followed their instinct and conviction of the worth of these texts which could – and did – so usefully pollinate Western literature. But the output of literature written in Persian in the huge geographic area called the Persianate world, stretching from Moghul India over Central Asia, to the Persian plateau all the way to Constantinople, is immense and is not limited to poetry. Scholars of Persian Literature form a tiny bunch of enthusiasts, toiling to embrace a mammouth thesaurus and labouring in our little niche to catch the interest of other literary fields. Our task is immense, or work force limited. We combine the isolation of Medieval Literary Critics with that of Persianists and the topic of the Darwin 2023 Lecture Series definitely resonates with us!

More generally, it is the universal human reaction to isolation which is central in the two medieval Persian tales discussed below. These passages examine the choice between isolation or friendship on the one hand, and isolation or marriage on the other. The texts were composed in the Persianate world during that exceptional early medieval, pre-Mongol period which develops despite three successive invasions. The period starts with the Arab invasion of the Persian empire, which is completed politically in 642 C.E. The Arab victory shatters the millennium-old rule of the Persian King of Kings and brings with it two remarkable elements: Muslim religion and Arabic language and poetry. The first literary example analysed here dates from these early days of the Islamic period.

Disputes within the Arab ruling families early on ripped apart the beautiful Islamic unity. Family members of Abbas, the uncle of the Prophet seize central power in 750 C.E., while the surviving Umayyad princeling

---

This article is based on two previous publications: C. van Ruymbeke (2016) *Kashefi's Anvar-e Sohayli. Rewriting Kalila and Dimna in Timurid Herat*, Brill: Leiden; and C. van Ruymbeke (2019), "What does Turandot want?", *Linee storiographiche e nuove prospettive di ricerca*, F. Bellion, E. Creazzo, A. Pioletti (eds), Roma, Soveria Mannelli (Catanzaro): Rubbetiino, pp. 269-290

flees to North Africa and will eventually establish the Umayyad dynasty of Spain. The new Abbasid dynasty leans on Persian clients and advisers. The Abbasid caliphs face the East, with its amazing cultural and commercial treasures. The new capital, Baghdad, is very close to Ctesiphon, the old Sasanian capital city. This is a signal confirming Persian culture's rebirth after a long occultation. Al-Ma'mun, the half-Persian son of caliph Harun al-Rashid (of the 1001 Nights), initiates a watershed translation movement of Greek, Hebrew, Persian and Indian, mainly scientific and philosophical, works into Arabic. This hybridity is at the core of the medieval Islamic culture's lustre. Arabic becomes the dominant lingua franca for the next three centuries. In the Persian lands, basing ourselves on surviving texts and historical chronicles, everybody writes in Arabic, whatever their mother tongue. Abbasid culture is radiant. As often, when culture flourishes, politics suffers. Indeed, the Abbasid power from its centre in present-day Iraq, can no longer reach into the most distant corners of the empire. In the 10th century, comes a revival of quasi-independent Persian princely houses in the far-eastern provinces of the empire.

This independence from the Arabic caliphate fosters the production of literary, especially poetical, texts written in Persian, for patrons who are of Persian descent and rule independently, paying homage to Baghdad in name only. We discover a transformed Persian language that quickly spreads westwards: while keeping its indo-european grammar and vocabulary, classical Persian has adopted the Arabic script and, depending on how close it is to the Abbasid central regions, it welcomes Arabic loanwords. Literature shows its patrons' interest for imperial Iran's greatness. While the Arab caliphs are ruling in name only, Iranian vizier families hold power in Baghdad.

This Arabo-Persian world is then overwhelmed by successive Turkish tribes. These new rulers are already converted to Islam and are content to take over political and military power, respecting the Arab Caliphs as heads of the Muslim community. The Ghaznevid Turks, followed by the Saljuq Turks, establish their rule over parts of the empire in the 11th and 12th centuries C.E. The Saljuq act as military governors, Sultans, but make allegiance to the Abbasid Caliph, who holds religious power. The Saljuq sultans patronise Persian culture and especially Persian Literature. For the next two centuries, the Easter Islamic world knows a golden age. The second literary example discussed below dates from the late years of the Saljuq era. A few decades later, the mind-boggling Mongol Invasions of the early thirteenth century come with terrible, comprehensive obliteration of this splendid golden age: human slaughter, destruction of towns and agriculture and destruction of learning and libraries. Islam itself almost disappears. It will quickly be reborn, as does Persianate culture. Religion and culture find a new vitality, but subtly different from before the Mongol invasions. Literature turns over a new leaf, changes its topics from philosophical to historical; and mystical texts thrive, with great poets such as Mawlana Rumi.

At the core of mystic teachings, the path towards the goal of ultimate annihilation within divinity starts with the voluntary search for solitude, for isolation from society, friends and loved ones. The reason for this rejection is that society and human affection are as unreliable as life itself. Our affection for other human beings is doomed to be betrayed at some point, certainly by death which finally will put an end to all human relationships. Instead of courting betrayal and its ensuing devastation, the mystic cleanses his life from unreliable, twisted human affection, in search of eternity and trust in God. Choosing total isolation from society, religion, learning, he puts all hope and energy in divine affection. The love God has for its creation makes the mystic secure that his affection is returned hundredfold, for all eternity. Nevertheless, most mystic seekers need to rely on a teacher, a "pir", a "sheikh" who will lead them at each of the stages of the difficult path to divinity. This relationship is the only exception for the "murid", the student, who lives in his self-imposed solitude, hoping to lose himself in the all-encompassing eternity of divinity.

But the main leg of pre-Mongol classical Persian literature are the works that cater for the paying patrons' down-to-earth interests. This is sophisticated literature, which addresses deadly serious topics, but also reflects the audience's joie de vivre and wide-ranging cultural interests. Poets compose complex bravura pieces praising their patrons. Authors write long narratives about famous pseudo-historical legendary figures, such as Eskandar (Alexander the Great) or pre-Islamic Kings of Kings of the great Iranian dynasties. There is little anxiety of influence; rewriting lies at the core of manuscript literature, in Muslim Persia as in the Christian West. There is no purist concern about an Ur-version, scribes are intervening and interfering in their source text, no two manuscript copies are alike. The authors' freedom resonates in their readers' freedom. These medieval narratives are open to interpretation, they address pro-active, imaginative readers, ready to dialogue with the authors. Under the guise of flimsy, sometimes farcical narratives, the most interesting, universal philosophical demonstrations are lurking. The authors propose a situation, provokingly dotting a few elusive clues within the stories, the readers must decode these, and satisfaction is immense when suddenly they have an epiphany and blinding light reigns. This is about leisurely, solitary, imaginative, slow, close reading; these are universal works that do resonate with the varied life experiences of each of the readers.

And so with the two narratives discussed here. They are of different types and genres, but they both have at their core a situation of isolation, and they deal with the danger of giving in to the lure of friendship or love relationships in order to escape it. In my first example, the hero is desperate to escape isolation, in the second example, the heroine prefers solitude over a miserable relationship.

The book of Kalila and Dimna

My first text is one of the most important Mirror for Princes of the late classical and medieval worlds: *The Book of Kalila and Dimna* (KD) This book of advice for would-be rulers, for rulers in place and for all those that are in close contact with the top of the political ladder, has been a true best-seller from antiquity up to the early-modern world. It is written as a collection of fables, with human and non-human animals who tell each other stories in which the characters tell each other stories and so on, thus constituting a remarkable example of embedded narratives or frame stories. The book's origin is lost in the fog of times. It is mostly presented as originating in India, but there are arguments against the existence of this mysterious eastern text, which has never been found. This Sanskrit legendary mother text is then said to have been translated into Middle-Persian (Pahlavi) in the fifth century C.E. at the Sasanian court. This text as well is lost. It might be that there was only one unique manuscript which was kept in the kings' treasury. Indeed, KD is sometimes described as a dangerous text, and I agree with this. The Abbasid caliphs managed to access it and had it translated into Arabic. This 8th-century Arabic text (now lost) survives in hundreds of manuscripts, the oldest of which dates back to the 13th century, 500 years after it was first written in Arabic. Each manuscript almost presents a different version of the Arabic text: the genealogical tree of the Arabic KD versions is extremely complex. Meanwhile the Arabic text was translated into Hebrew, from there into Latin and Greek, and from there in all imaginable languages, Italian, Spanish, German, French, English, ... The Arabic original was also translated and rewritten at least seven times in Persian throughout the middle ages and pre-modern times. Some Persian versions are in prose, some in verse and some in prosimetrum, this hybrid form mingling prose passages with verse inclusions. This is the case with the specific version used here, written for Turkish rulers in the 12th century, by the court administrator, Nasrollah Monshi. It claims to be a direct translation from the 8th-century Arabic version. In his prologue, the author explains that his audience's knowledge of Arabic was no longer sufficient to read in Arabic and people were too lazy to run to a dictionary to find the meaning of every other word in the text. A Persian translation was indispensable as this fascinating work ran the risk of completely disappearing from memory.

There are three important points that need to be made in regard to reading and understanding KD. First: contrary to what is often repeated, fables are open to interpretation; like diamonds, each of its facets is meaningful, what happens to each character in the story is important and should be considered on its own. So, when an author imposes a "moral lesson" on a fable, he artificially and tyrannically seeks to manipulate us readers to view the story in a single manner, thus seeking to manipulate us. Second: KD addresses kings, important, busy people who cannot waste time reading silly stories. The success of this book copied in often deluxe manuscripts throughout centuries across cultures, languages and religions, is a clue that its advice is not only timeless, but also extremely useful for people in positions of power. Third: the fashion of writing children books is not born yet in the medieval period and although nowadays the stories of KD in the Arabo-Persian world are stories read in primary school, this comes from the misunderstanding that fables are charming children stories. On the contrary, KD is a grim book, deadly serious and indispensable.

One of KD's most important messages, which is hammered home time and again, is that the person at the top is isolated, cannot trust anyone, especially not those attempting to ingratiate themselves with the plan to manipulate the ruler by pretending friendship or love. In general however, on religious grounds, solitude rates far below the companionship of wise and good people; God has made men mutually interdependent and has shaped civic life.

Christine van Ruymbeke

> *Several religious leaders and masters of the faith [...] have said that
> the society of a good companion is better than solitude, but when an
> amiable friend is not to be had, solitude is better than society.[...]
> from the purport of the tradition.* **There is no monasticism in Islam'**
> *it is understood that the advantages of society are superior to the
> utilities of solitude.*

But, warns KD, a ruler should resign himself to the knowledge that he will
attain neither ideal friendship, nor ideal love. Truly, he can never trust anyone, whether friend or lover: he is condemned to the harshest of solitudes,
alone at the top. Friends and lovers manipulate and even betray. What is
more, his rule will suffer and he will not do his duty towards his people. On
this note, it is not recommended either for the ruler to transform solitude in
mysticism. Searching to attain purity and living as a hermit is not an answer:
the business of being a king is about active involvement in society.

If friendship and love are dangerous for the unhappy lonely ruler on the
receiving end, the good news is that these two feelings, as well as flattery or
even betrayal of the given word are all powerful manipulative tools which
the ruler should harness to further his own political ends.

The KD's self-other relations illustrate the clash between the two accep-
tances of the "politics of friendship," already perceived and studied in Antiquity. The safe sort of friendship in this case is the one we would view as
cynical, the one influenced by the reciprocity model of a finite and politi-
cised *homonoia* (concord). This type of "friendship" does not arise from
completeness, it arises from insufficiency and it is marked specifically by
the need for reciprocity. This is a useful relationship that follows the same
rules as, practical and political alliances. The bonding ends when so dic-
tated by the changing necessities of pragmatic politics. This can be a useful
friendship between two actors who understand the rules of the game and
temporarily join interests, without being sentimentally involved; or it is the
relation established between a manipulator and a victim; the latter will end
up broken and bitter because he misunderstood interested closeness for sen-
timental friendship.

This is opposed to the feeling we call friendship nowadays: the romanti-
cised friendship. This is a rosy view of *consensio* (communion) and *philia*
(love). It is a bond of gratuitous asymmetry and infinity, one that is strictly
private and that cannot be translated into the political sphere. It is understood as an almost sacred attachment relating to feelings of love, completeness, joy and sharing fun. Medieval and pre-modern authors are aware of its

existence, but warn that it is incredibly rare, and mostly traitorous, created by wishful thinking. It should only be entered into after mature examination and test periods as to the virtue of both parties. When such an exceptional bond is shaped, between two equally excellent and virtuous partners, it is the vector of ineffable and infinite delights. Human beings have a great longing for it and prefer to ignore danger signals, in order to keep alive their pipe dream of shared and virtuous amity. It is a psychological escape from the fundamental human dislike of solitude coupled to a need for recognition and sense of belonging.

The *KD* authors saw this as one of the critical perils a ruler has to ward off: a weakness on which unscrupulous, ambitious tricksters will play. We find ubiquitous warning flashes about friendship throughout the text. Most importantly, KD lavishes a lot of attention on the choice of the persons who should be allowed to enter into a friendly relation. Perhaps the cruellest example of the necessary loneliness of the ruler and the danger of fickle likes that are too hastily transformed into the trust that should only be given to the true friend are expressed in the grim tale of the Monkey and the Tortoise.

The old Monkey-king is banished from his tribe. Forced to become an unsatisfied hermit, he lives on the shore, in a fig tree, waiting for death's call. He eats figs and his boredom is great; his only fun is the sound of the figs when he throws them in the water. He is unaware that he feeds a Turtle swimming close by who receives these figs as incredible bounty. The Turtle misunderstands the Monkey's ennui for generosity and decides that he cannot miss his chance with this largesse-scattering creature. He approaches the Monkey and initiates friendship. The Monkey misunderstands the Turtle for a soul-mate who approaches him to discourse on philosophy – too delighted to finally have someone he can talk to. He thus talks away, while the Turtle makes intelligent noises but in fact really gobbles the figs. They spend much time in separately satisfying intercourse. Meanwhile, the Turtle's wife worries at her partner's neglectful absence and doesn't believe his story of the figs and the chatty Monkey. Deeply jealous – Hell hath no fury etc. – she tries a cruel counter-move: she pretends to be ill and that she can only be cured by a monkey-heart. The Turtle, though hesitant, agrees to invite the Monkey for lunch, when the Turtle couple will butcher him to extract his heart. Delighted with this attractive social invitation, the unsuspecting Monkey hops on the Turtle's back to cross the water. Underway, the chatty Turtle explains the wife's illness and the importance of the heart cure. The Monkey has a true "Hitchcock moment" as it suddenly dawns on him that he can't swim and that he finds himself in the middle of the water, sitting on the back of a dangerous though charming psychopath who calmly explains that the plan is to rip out the Monkey's heart in order to feed it to his ailing wife. This is one of the most charged moments in the whole book and the Monkey's iron self-control and masterly psychological manipulation are truly inspiring. He avoids contradicting the psychopath: he agrees that the cure is common knowledge and pretends he will help. Next, he introduces the trump card that "he left the tree without his heart and they need to go back to fetch it" and manages to make the Turtle swim back. Once safely up the fig tree again, the Monkey wipes his brow and, delighting in his wonderfully safe isolation, he boasts that he will not be taken in twice. The bewildered Turtle is left to bemoan the loss of a generous friend.

The two ill-assorted mates failed to realise that their feelings are not on the same level. The Turtle is interested only in material gain: he establishes an interested relationship in order to continue enjoying the figs. Obtaining the figs has been so easy that he pushes further: it is just one little step further from a fig to a heart. The mistake the Turtle makes is to remain unaware that this little step for him is in fact a huge jump for the Monkey who is bound to protest. Also, the Turtle does not account for the fact that he is planning to kill the Monkey who throws the golden figs. The Monkey, for his part, has not carefully vetted this new person, in his joy to find someone to speak to. So bored is he that his most intimate, soul-searching philosophical thoughts pour out of him and there is no need for a reply from the Turtle. However, he trusts this unknown person with his life, across the water, putting himself fully at the mercy of someone he has not vetted carefully. But the deepest ad hoc who reads the KD, is that our old ex-ruler Monkey is unable to understand that even when he no longer has power and gold to distribute, yet he still has things for which others may be ready to pretend friendship. The Monkey forgot rule number 1 which is the necessity for a ruler to avoid entering in a friendship relationship. The story shows that this danger alas, does not end with retirement.

Haft Paykar

Nezami Ganjavi is one of the giants of Medieval Persian poetry. He is a challenging poet, a fascinating word-smith and a baffling narrator. He lived in present-day Azerbaijan, at the end of the Saljuq era, and worked on commission from Turkic governors in the region. His five long narrative poems are famous and baffling. His most intriguing work is called Haft Paykar

(HP), usually translated as the Seven Beauties. It is the tale based on the life of a young prince and later Sasanian emperor of Iran, referring to a historical figure, Bahram 5th, nicknamed "Gur", who had ruled from 421 to 438 C.E. The most celebrated moment in the HP, is an astronomical love journey. King Bahram has claimed the daughters of seven neighbouring rulers. The seven regions of the ancient world were ruled over by one of the known planets. Each of the planets was associated to a day of the week. Nezami also relates each planet to a colour. So, each of these women come from one of the seven regions, each ruled over by a planet. Each bride lives in a separate pavilion which is coloured according to the related planet, and on the given day of the week, the king, dressed in the relevant colour, spends time with that bride, who tells him a tale. The series starts on Saturday, when Bahram, dressed all in black, visits the daughter of the king of India, in the Black pavilion, ruled by Saturn. It is usually accepted that these seven women are teaching Bahram some essential wisdom related to his journey to become a perfect king and perfect man. In my view, however, the stories and their content have not yet been satisfactorily explained and decoded.

The episode I want to develop here is placed at the very centre of the astronomical week, when King Bahram, on the Tuesday, visits the Red Pavilion, ruled over by the red planet Mars and he hears from the daughter of the Slav king, a story which never stops fascinating us. It is the story which will be chosen by Puccini for his early 20th-century opera Turandot.

The story is thus: An aged king has an only daughter, very beautiful, very intelligent and extremely well educated, a true blue stocking. Marriage candidates present themselves. The father is anxious but she is adamant that she will only marry someone of her choice. Nezami, I believe only half-teasingly, comments on marriage as a demeaning status for someone as accomplished as she is: "How can one so unique agree to marry?". The father bows to his beloved daughter's whim. Clearly, she is not averse *per se* to marriage, but she just wants to find the right person. The princess builds herself an isolated fortress in the mountains. Its entrance is hidden by magic calculations. The path to the fortress is guarded by magic killer robots that exterminate any traveller on the road. Before repairing to the fortress, she paints a beautiful self-portrait and has it plastered on the city's door with the accompanying text that he who wants to win her must fill four conditions: (1) have a noble name and physical beauty; (2) neutralise the killer robots; (3) find the hidden entrance to the fortress, but refrain from entering it; and (4) answer questions in a satisfying manner. The winner will become her husband and inherit the father's throne; the loser will lose his head.

Christine van Ruymbeke

The challenge plastered on the door of the city is so enticing that soon enough the walls of the city are covered with the heads of the unsuccessful candidates. She has no pity for those poor losers. It seems evident that if they were stupid enough to take up a superhuman challenge without being forced to do so, if their passion ruled their intelligence, they were obviously unsuitable.

Until the story's hero passes the town one day and, as all the others, falls in love with her portrait and decides to take up the challenge. But, since he is also wise, he first goes to a hermit who teaches him all the princess knows. Once ready, he enters his name and starts out towards the fortress. He fulfils the first condition of pedigree and physical beauty, next he manages the second test by annihilating the robots, next he can work out where the hidden entrance to the fortress lies. The princess remains hidden and orders him to return to town and await the riddle test. What does the princess want? The nature of the challenges sheds light on her hopes: we already know that she wants someone of good pedigree and physically attractive. She also needs someone who is ambitious enough to enter into the contest and courageous enough to fight against the killer robots and he needs to be clever enough to defeat them. She wants someone intelligent with knowledge of architecture, numbers, mathematics to understand how to find the entrance. Someone who has enough self-control not to rush into the fortress, but who will submit to the final tests. So far, we can indeed only agree with her expectations for an ideal husband and ruler at her side.

The final test is the core of the story: the riddles will show whether the hero is ideal husband material. It is important to note that Nezami does not give us the solution to the riddles, but he teases us by giving us a false solution, so openly silly and pointless that *it is impossible to take seriously*. It has nevertheless mostly been adopted as an anti-climactic, puzzling solution. Riddles are frequent in Medieval narrative contexts, where they take the form of metaphoric questions. Riddling questions are usually put to a male character, very often in a marriage context. Frequently associated with death and related to the Freudian archetype opposing the life forces of Libido and Eros, to Thanatos, the death instinct, they create tense moments in a story. Rather than testing "intelligence", they ascertain the respondent's religious or philosophical views and thus his suitability to become part of a new social structure. Their lightweight punning structure and their irrelevance are jarring compared to the punishment attached to failure. I argue here that the interpretation of such episodes might benefit from an analysis that does not lose sight of the humorous essence of riddling as a literary device.

The princess's riddles are a compound of several recognised riddle genres. They are true metaphoric riddles in that the cluster of questions contain enough elements for the riddle to decode the metaphors. They are also wisdom questions, as the answers need a thorough grounding in numerology and poetical metaphors. They are sexual riddles as they partly refer to physical lovemaking. Nezami adds an additional difficulty by making this a silent riddle match. Not a word is expressed by the princess and the candidate; they are also unable to see each other. Thus, the candidate has first to decode what the question is, before answering in kind. We, as readers follow the silent match, doubly puzzled by the princess's questions and the hero's answers. The hero shows that he is equal to the difficult match, while we remain baffled.

The first question: The princess sends two tiny pearls, which the suitor weighs in his hand. His response is the addition of three similar tiny pearls, weighing the same, making five in total.

The second question: The princess now grinds the five pearls to dust, weighs this powder and mixes it with sugar. The suitor adds milk, in which the sugar dissolves. Having drunk the sweet milk, the princess makes a paste with the dregs of pearl-dust, which she weighs to check that the full weight of the five pearls is still there.

The third question: the princess sends one of her rings. The suitor keeps the ring and sends back a priceless un-pierced pearl.

The fourth question: the princess finds in her necklace a pearl exactly similar, binds the two magnificent jewels together. The suitor adds a blue glass bead without any value because he is aware that it is impossible to find a third matching pearl.

The princess laughs, wears the two pearls on her ears, the bead on her hand, and agrees to the match. Her father is baffled. Like us, has only seen the silent exchange between the princess and the candidate. He has not understood a thing. The pseudo-key now provided by the princess has often simply been adopted as the bona fide key, although at first blush, it is unlikely that Nezami would have provided the explanation to his wonderful riddles, thus deflating his audience's joy in the search to unravel the learned enigma. There is a tradition of providing pseudo-keys, particularly frequent with sexual riddles. It is a game, meant to surprise and amuse the participants, disorienting the respondent, but often providing essential indications ad absurdum that ultimately help unravel the riddles. The pseudo-solution plays on the ambiguous imagery and proposes an innocent "wrong" non-erotic explanation to an erotically charged riddle. The passage also makes excellent storytelling. It provides an amusing narrative detail showing that the Princess is unwilling to explain the riddles' charge of eroticism to her parent. Here is how she explains the Q&A:

> *I loosened the pearls from my ears, By their example I said "Know that this life is but two days" – He adding three to two, replied: "It passes quickly, even if it lasted five days". I ground the pearls and sugar to dust, saying: "This life is mixed with desire. Who, by spells or by alchemy, can separate the two?" He poured the milk, so the sugar dissolved, the pearl dust remained, saying: "Sugar mixed with pearls will melt and vanish with a drop of milk". In drinking from his cup, I signed to him that I was but a child to him and I sent the ring consenting to marry him. He sent me back that pearl, "Like this, you'll find no mate for me". Joining my pearl to his, I showed that I was his equal. He saw that there was no third pearl like these two in the world, and added a blue bead against the Evil Eye. When I put on that bead, I meant thereby, the seal of my consent: Upon my breast his seal of love is that which guards my treasure-hoard. Because of those five mysteries, I bowed before his mastery."*

Nezami flashes several signals that this is a pseudo-explanation, but he also provides clues to the real meaning of the riddles. One indication that this is not the full explanation is that only part of the exchange is explained, for instance no mention is made of the repeated weighing of the pearls. The jumble of elements contradict each other, and contradict the context of the riddles. The opening gambit relates to the stock remarks of sages and especially mystics who are aware that life passes in the blink of an eye and that this short time ought to be used in preparing for the here-after. In such a frame of mind, the aspiring mystic is encouraged to stay away from human entanglements and human love affairs and to focus on searching for divine love. Such an opening could hardly lead to a marriage proposal. The numerals two, three and five have no relevance in this con-text, but, in stressing them, Nezami teasingly highlights their importance as a real key to the riddles. Continuing with the pseudo-mystical theme, the Princess regrets that desire (sugar) is mixed with life (pearl dust). While Nezami gives a correct clue with the sugar-sensuality metaphor, the nega-tive way in which it is presented here is again counterintuitive in a marriage context. The Princess then abandons the mystical ambiance and states that only magic can separate life and desire. But the suitor easily separates them in milk. Milk often appears in sexual metaphors, but the Princess leads us astray in explaining that the suitor opposes it to maternity (breastfeeding stops sexual desire). The suitor contradicts himself with the conventional view that the point of marriage is procreation, opposed by the decision to stop sexual desire within married life. He nevertheless provides a clue to the erotic ambiance of the dialogue. The trick with the milk appears to convince the Princess, but she misunderstands her role in procreation. She admits to being like an infant (drinking milk) towards the suitor. The infant reference re-introduces the mystical ambiance of the disciple who feels as a child towards his pir (teacher). Evidently, this sudden self-belittlement and pledge of filial obeisance to her future husband are opposed to the Princess's independent character. Furthermore, it does not make sense as, by drinking the milk, the Princess is in fact nurtured with the desire that was dissolved in the milk. Next, she offers the suitor a ring, which signals her agreement to the match. This is absurd, as this should immediately close the exchange and express the suitor's victory. But the suitor reopens the dialogue on an aggressively self-satisfied tone, sending a unique priceless pearl, meant to advertise his own unique priceless worth. Recovering some of her spirit, the Princess contradicts him with a similar pearl, stating that he and she are of equal worth and both unique. This is opposed to her recent recognition of male mastery. The suitor adds a bead to protect them both from the evil eye. Again, this is a clue to the real explanation. Mixing real and fake elements, the Princess concludes that she wears the modest bead as a signal that her treasure-hoard belongs to and is guarded by the suitor. Intriguingly, while there should be four riddles, here she concludes with the mention of five mysteries.

Has Nezami given us enough clues to work out what has really happened? How can we understand the silent dialogue? It is apposite here to refer to the isolation of the reader challenged by the medieval author to work out pro-actively the meaning of the narrative. To start with, I based myself on two premises: (1) the exchange is of paramount importance for the princess in her search for an ideal husband: it should target qualities which the candidate has not yet demonstrated; (2) all the elements in the riddle ought to be meaningful in our decoding effort. These elements are: pearls, sugar, milk, ring, blue bead and the numbers 2, 3, 5. The weighing of the pearl dust must also be meaningful. Once I understood the meaning of these numbers, and the metaphor of the pearl, the pieces of the puzzle fell easily in place.

The first riddle opens on a question which consists in two tiny pearls. In Pythagorean number lore, the gendering of numbers is common. Female numbers are even, odd numbers are male. So, the princess sends two pearls, in essence announcing: I am a woman. The hero sends them back with an additional three similar tiny pearls, saying: I am male. In so doing, he forms the number five, the conjunction of male and female. His answer is also a question: "Well, what about it?"

The second riddle: the princess answers by grinding the five pearls to dust, thus reducing male and female pearls to atoms mixed together: this is an evident metaphor for making love. She weighs this powder and adds sugar: the sweetness of lovemaking. The hero dissolves the sugar in milk and the princess drinks the sweet milk, symbolically tasting this love-making. But when she weighs the pearl powder again, it is still the same as before the love-making. Nothing was lost but nothing gained either.

The third riddle: she sends a ring, a traditional reward for sexual favours. This is also probably a goodbye gesture to the hero, as the love-making has produced no lasting gains. Now comes the most important moment of the riddle test. The suitor needs to re-open the dialogue. He has brought with him a peerless, unpierced pearl. The unpierced pearl is a well-known metaphor in Persian literature, for the maiden, the virgin girl. The suitor is protesting that even without involving sex, the princess is peerless and priceless.

The fourth riddle: the princess finds a similar pearl in her necklace, – thus a pierced pearl –, and binds it together with the unpierced one. She is again forming the number two of the beginning, the female number. She is telling him that she might well be priceless when unmarried, but once pierced, once a wife, she will stay exactly the same: priceless and unique. Her unspoken question is evident: how will the suitor act with her once he has become the husband, the master?

His answer is perfect: he adds a blue glass bead, which has no value at all, but is used as protection against the evil eye. By adding the bead to the two wonderful pearls, he forms the number three, the male number of the beginning. He tells her that his male identity has no other value and no other aim than protecting her peerless value as virgin and as wife, against the evil eye.

And naturally, the princess is delighted and agrees to marry the one who understands that his role as husband is to protect her, not to enslave or to change her in any way.

The pseudo-key presents her as being overwhelmed by his superiority after the trick with the milk. Her self-belittlement and his self-satisfaction are to be put in context with Nezami's representation of the father's male dog-matism. He interprets the conclusion of the riddle-match as his daughter's bowing to a dominant master: "When the king saw that wild colt tamed by wisdom's whip." He views the match as one of submission and subjugation towards a strong male who will become the husband. In his eyes, this marriage is not about an equal match which would safeguard his daughter's independence and power. Beyond humour, the pseudo-key's exaggerations offer Nezami the opportunity to highlight the almost revolutionary feminism of his story.

The suitor manages to express with his blue glass bead that power and control are not an issue because male power exists in order to protect female independence and perfection. Nezami's point about the conundrum of wise women accepting to give up their independence and ensuing solitude, is that it can only be to men of equal wisdom. But the reader is left with a nagging doubt. Considering that at this stage in their acquaintance, we cannot yet speak of a relationship of true love, are we then dealing with manipulation? And in that case, who is the winner of the match? The princess or the suitor?

Christine van Ruymbeke

Further reading

Morgan, D. (1988), *Medieval Persia 1040-1797*, Routledge Publishers, London and New York.

Nizami (1995), *Haft Paykar, a Medieval Persian Romance,* translated by J. S. Meisami, 307 pp., Oxford University Press, Oxford, New York.

van Ruymbeke, C. (2012), 'Murder in the forest. Celebrating misreadings and rewritings of the Kalila-Dimna Tale of the Lion and the Hare', *Studia Iranica* **41**, 203–254.

van Ruymbeke, C. (2016), *Kashefi's Anvar-e Sohayli, Rewriting Kalila and Dimna in Timurid Herat*, 426 pp. Brill, Leiden.

van Ruymbeke, C. (2019), 'What does Turandot want? From Puccini's Freudian riddles back to Nezami's silent Pythagorean questions', *academia Nazionale de Lincei Atti sul Medioevo Romanzo e Orientale* **XI Colloquio**, 191–211.

2 Isolation of asylum seekers: immigration detention in Australia

AMY NETHERY[1]
Deakin University

Abstract: Australia's policy of mandatory, indefinite and unreviewable immigration detention was introduced in the early 1990s to respond to the arrival of asylum seekers by boat. The policy persists despite its failure to deliver policy goals, vast expense, international condemnation, and human damage. What explains this persistence? In this essay, I argue that immigration detention is best understood as the most recent iteration of administrative detention, a form of non-judicial incarceration with a long history. Governments in settler colonial Australia have found administrative detention indispensable for classifying and then incarcerating groups of people regarded as a threat to national security or identity. Significant historical examples include Aboriginal reserves, quarantine, and enemy alien internment; today's offshore and onshore immigration detention centres share a similar purpose and character. Sites of unmitigated executive control, these different forms of administrative detention are control regimes with punitive effects. By demonstrating the embeddedness of this form of governance in Australia, the essay provides an endogenous explanation for the persistence of immigration detention, despite its harms.

INTRODUCTION

Mehdi Ali was 15 years old when he travelled by boat to Australia seeking asylum in 2013. He was born in Iran to a family who are members of the Ahwazi Arab community, a group Amnesty International has identified as

---

[1] This essay draws on themes and research published elsewhere, including (Nethery & Holman 2016, Nethery 2021).

persistently marginalised. Fearing persecution, Mehdi made the journey to Australia to seek asylum. He travelled to Indonesia alone, where by coincidence he found himself smuggled onto a boat with his cousin, Adnan, aged 16. The pair became inseparable, and it was a good thing, because Mehdi and Adnan were to spend the next nine years in Australia's immigration detention system.

For the first nine months of his incarceration, Mehdi was held in a detention centre on Christmas Island, which has been part of Australia's detention network since 2005. Some 2,600 kilometres from the nearest capital city, Perth, Christmas Island is Australian territory but geographically part of the Indonesian archipelago. It is home to the extraordinary phenomenon of the annual red crab migration, which Sir David Attenborough has described as "a great scarlet curtain moving down the cliffs and rocks towards the sea" (Parks Australia 2001) . Mehdi, detained in a high-security detention facility with double-fencing, metal gates with double-airlocks, security cameras and isolation cells, did not get to see the red crabs.

After nine months on Christmas Island, Mehdi and Adnan, now aged 16 and 17, were transferred to one of Australia's two offshore processing centres. Not yet adults, they escaped being sent to Manus Island in Papua New Guinea, which hosted only men. Instead they were sent to Nauru, a small Pacific island 4,000 kilometres from Sydney. Mehdi was housed in a family camp, and allowed to attend the detention centre school. Guards referred to Mehdi not by his name, but by his boat number: ANA020.

Mehdi's refugee status was recognised in 2014 while on Nauru. He, cousin Adnan, and the 28 other unaccompanied minors were released from the detention facility to community accommodation, but it was not safe living in the local community. There was antagonism from local Nauruans, who did not want the refugees on their island and saw them as a privileged group receiving Australian money. The foreigners were subject to a spate of violent attacks, including beatings, robberies and rape. The Nauruan authorities were not concerned about these attacks, and no one has been charged for the crimes (Human Rights Watch 2016).

Mehdi was 17 when he watched his friend and fellow refugee Omid Masoumali set himself alight during a visit from United Nations officials. Omid remained on Nauru for 24 hours until he was taken to an Australian specialist burns hospital for treatment. He died the following day (Doherty 2021).

Mehdi and others protested the authorities' slow response in treating Omid. He was jailed, and there he was stripped naked and beaten, and eventually released without charge. He started having anxiety attacks, and developed insomnia (Mehdi 2022).

Meanwhile, back in Australia, a unique political crisis plunged the Coalition government into minority status, and some quick-footed independents used the opportunity to pass the so-called Medevac Bill (2019). The Bill created a pathway for critically ill people held in Australia's offshore detention regime to be transferred to Australia for medical treatment. The Minister and his department had been blocking or delaying such applications so that some applications were taking 2 to 3 years for approval. The Bill was short-lived, however, as the Coalition regained its majority status at the next election and the Medevac Bill was repealed within eight months.

Mehdi, diagnosed with post-traumatic stress disorder, was one of 192 refugees brought to Australia in 2019 under the Medevac Bill (Kaldor Centre 2021). Over the next few years he was transferred six times between hotels and detention facilities in two cities. The transfers were sudden, often early in the morning, and in handcuffs (Mehdi 2022).

Finally, in 2021, he was held in the Park Hotel, in Melbourne, along with about 50 other detainees. The windows in the rooms had been screwed shut so detainees couldn't communicate with supporters outside, but also meant that there was no fresh air. Half the detainees caught Covid-19. In all that time, he didn't receive the medical treatment he needed.

In March 2022 Mehdi was granted a resettlement place in the United States. His cousin Adnan followed him a month later. During that same period, in the weeks leading up to the 2022 federal election, all remaining detainees in the Park Hotel were released.

Mehdi had entered Australia's system of immigration detention at the age of 15, and left when he was 24, Fig. 2.1. He experienced terrible conditions and abuse, violence, and active neglect which has proved devastating to the mental and physical health of everyone who has been subject to it.

His story gives us insight into Australia's onshore and offshore immigration detention systems, and the experiences of those who have been subject to it. His story is not unique. In fact, it is overwhelmingly similar to that of all 3127 people who sought asylum by boat between 2013 and 2014. All were sent offshore, with a promise that none would ever step foot on the Australian mainland because of the mode by which they travelled to Australia to seek asylum.

In 2021, half of these people were still detained. 17 people have died in the offshore immigration detention system: most of these deaths are by

Figure 2.1 The quote is from Mehdi Ali himself, prior to his release.

suicide or medical neglect (Monash University 2022). Many others have suffered permanent psychological damage. The cost of Australia's off-shore detention centres since 2013 has been estimated at AUD$ 13 billion [US$ 8.9b]. The private corporations tasked with detainees' care have made extraordinary profits. Nauru, bolstered by Australian cash, has experienced a rapid and dramatic decline towards authoritarianism. Meanwhile, refugees in the Asia-Pacific region are left without a pathway to permanent protection, and many thousands are stuck in Indonesia, a country that does not recognise them under international law.

How should we understand this system and the experiences of Mehdi Ali? I would argue that while a discrete analysis of Australia's tough approach to asylum seekers answers some questions about the policy, there is a lot it doesn't explain. This is because much of the policy responds to emotive, rather than rational stimuli, it is fiscally irresponsible, and disproportionate to the problem it tries to solve. The policy goals have been met partially or not at all, and the human harms have been significant. We can deduce, therefore, that there are deeper social and political functions at play.

The answer to this question comes from a much broader analysis of the role that incarceration has played in the settler colonial state of Australia. Australia, of course, was a convict colony, and the idea of incarceration, containment and control was at the heart of the whole enterprise. Thenceforth, both judicial and administrative forms of detention became deeply embedded within the colony. Both forms of incarceration quickly became indispensable tools for establishing the colonies, and later the new Australian nation. Administrative detention has proved particularly useful for settler colonial governments in Australia because it has enabled them to classify and then incarcerate and control certain groups of people. Its effectiveness at this task, and the way the policy provided governments with pockets of unmitigated executive control, meant that it was an attractive template for which policymakers reached when faced with biopolitical problems.

They were, and remain, especially useful in managing what are perceived to be problem populations. I argue that to understand the longevity of immigration detention, this contemporary form of administrative detention must be seen not as exceptional, but as a recent iteration of a much longer form of incarceration, one that has been central to shaping the character of Australia. To prosecute this argument, this essay first explains what is meant by administrative detention and its importance to colonial, and later Australian, governments before returning to provide a fuller picture of the policy of immigration detention.

Administrative detention: classification & executive control

Considering what it signifies, the term 'administrative detention' is less widely used than it should be. Administrative detention constitutes any form of non-criminal incarceration that is ordered and authorised by the executive arm of government, rather than the judiciary. As detention without trial, it is therefore constitutionally distinguished from imprisonment in the criminal justice system, even if substantially the distinction between these forms is not always obvious.

Modern forms of administrative detention emerged at the same time as the prison, and many varieties have been implemented throughout the Western world. The great historians of incarceration – David Rothman (1971), Michel Foucault (1967, 1977), Erving Goffman (1961) and others – have demonstrated how this form of incarceration became an indispensable method by which governments imposed order over their populations since the mid-1700s. Rothman's history of the evolution of incarceration in the United States, and Foucault's tracing of carceral sensibilities in Europe, provide a genealogy of this form of government control. Many historical forms, such as poorhouses, lunatic asylums, and leper colonies, were rendered unnecessary or socially unpalatable by the modern welfare state and developments in healthcare. But it is also true that many people who once would have been subject to these older forms of administrative detention are now caught up in the judicial prison system, ostensibly for criminal activity, but also for reasons of poverty, race or mental illness (Fassin 2017).

In liberal democracies, administrative detention has a specific constitutional status. The ancient principle of *habeas corpus* sits at the heart of modern constitutions and acts as a preventative for arbitrary incarceration, yet many liberal democracies, such as Australia, make exceptions for non-citizens. To understand how administrative detention works, it is useful to compare its processes with those of judicial incarceration. We must proceed here with caution, because it would be folly to argue that the judicial prison system operates according to its fair, rational and 'blind' ideal: it clearly does not. Nevertheless, comparing processes between prisons and administrative detention – even in terms of the ideal processes – can shed light on how the latter works. People subject to administrative detention are detained not because of something they have done, but because they have been classified into a social category whose detention is required either by parliament or executive order. In some specific cases, this means people are detained simply because of who they are. Laws that require the administrative detention of people according to their social category result in incarceration of whole groups of people at once. There is no weighing of evidence, nor investigation into whether detention is correct, justified, or proportionate in each instance.

Administrative detention is also unique in that in most cases, detention is indefinite. People remain detained until their circumstances change and they no longer fit that particular administrative category, an executive order grants the whole group released, or the detainee dies (Thwaites 2014). Finally, administrative detention is characterised by unmitigated executive control. Throughout history the different forms of administrative detention have been constructed so that they are subject to minimal independent oversight and judicial review.

The power of incarceration to manage populations means that administrative detention is attractive to governments of all ideologies, historical and contemporary. In contemporary Western liberal democracies, sites of administrative detention are upheld by a legal fiction that they are not punitive in intent, and therefore do not breach the principle of the separation of powers. Yet the lack of regulation, transparency and accountability over daily life within these closed sites means that rights cannot be protected, and neglect, abuse and general violence can and does occur.

This is not to say that sites of administrative detention are necessarily 'out of control'. Instead, it is truer to say that the level of control is left entirely to the discretion of detention authorities. Authorities may choose to neglect a sick detainee, or to micromanage when they can use the bathroom or how they might spend their days, or inflict abuse or violence. This arbitrary and often excessive control is crucial to the experiences of detainees, and therefore the meaning of detention (Bosworth 2014).

What is created are unregulated control regimes that are punitive in effect and damaging to the people subject to them. In their mildest forms, sites of administrative detention are temporary holding zones. More serious forms resemble high-security prisons, designed to harm and punish the people detained. At their worst, forms of administrative detention have been deployed as part of the architecture of genocide.

Administrative detention in Australia

In 1788, the land that became known as Australia was claimed by the British Empire as a convict colony. The ideas of incarceration, containment and control sat at the heart of the enterprise. Within the first 50 years of the colony's establishment, several different forms of administrative detention emerged, each one to deal with a particular biopolitical challenge. These different forms of administrative detention have proved particularly useful for settler colonial governments in Australia because it has enabled them to classify, incarcerate and control certain groups of people. And so administrative detention has formed an institutional template for which policymakers reached time and again when faced with biopolitical problems.

The first system of administrative detention to emerge in the new colonies was quarantine. From the 1830s, Australia began to establish a system for quarantining new arrivals to the country. At first, people were held on board their ships in the harbour, but in the 1850s the colonies began to construct quarantine stations at strategic locations. From then until the 1950s, when air travel made it unfeasible, quarantine was an unnegotiable part of the process of entering the country. This was a full 100 years after the UK and other European states ceased the practice. Quarantine policy was maintained in Australia because it worked: being so far from most points of departure, colonial governments were able to keep Australia largely free of disease. Historians have argued that an additional attraction of the policy was the social, rather than medical, benefits it provided colonial governments. Quarantine played an important role in the imagining of the new nation of white, clean, strong, healthy citizens. It also played a role in classifying the population at the very point at which they entered Australia: separated into class and racial categories, people experienced their entry onto Australian shores in vastly different ways, communicating a clear message about who was welcome, who belonged, and who was merely tolerated.

Around the same time as quarantine was established, another form of administrative detention was taking shape. A network of protectorates, missions and reserves emerged for Australia's First Nations people. Initially, these were designed with the charitable, albeit misguided, intent to protect people from the violence of the frontier wars, and to 'smooth the dying pillow' of a race of people it was believed were destined for extinction (Reynolds 1989). No regard was given to which nation they were from: all Indigenous people were treated as though they were the same, and the reserves were often far from family and country. All aspects of their lives were controlled by the Chief Protector: from the tasks that filled their days to who they could marry. Living under the Aboriginal Protection Act became increasingly tougher around Australia's federation in 1901. Around this time it became clear that First Nations people were not, in fact, dying out, but instead the population was becoming more complex with the birth of children with lighter skin. So the reserves were reimagined again, and this time they played a central role in the policy of removing children of mixed parentage from their families, in a policy that has since become known as the Stolen Generations (Wilson 1997). These children were permanently taken from their families, banned from speaking their language or engaging in any cultural practice, and trained for domestic work or labouring jobs in white households. While most reserves were dismantled in the 1960s and 70s, some were still operating until the mid-1980s.

The third major example of administrative detention was Australia's extensive system of enemy alien internment which detained thousands of people during both world wars. Australia was already a migrant nation by 1914, but people who were born in enemy nations, or people whose parents were born in enemy countries, were interned throughout both world wars. Most of those so-called 'enemy aliens' were interned on spurious grounds amounting to little more than hearsay, gossip and rumour (Saunders 1993). Because of its perceived enthusiasm for internment, Australia also interned thousands of people on behalf of allies Britain and the US. Most interns were released back into the community at the end of the war, but for some groups internment was a step toward permanent exclusion. This is the case for the small but long-standing Japanese community. Because of the White

Australia Policy of 1901, all people of Japanese heritage in Australia in the 1940s would have either been born in the country, or migrated over 40 years earlier. Nevertheless, every person of Japanese heritage was interned during WWII, and at the end of the war, deported without exception (Nagata 1996).

What can we learn from this survey of different forms of administrative detention? The first lesson is to avoid the temptation to treat them as exceptional and unique instances of incarceration. Instead, these three forms of incarceration must be understood as variants of the same: a distinct type of incarceration with a particular social and political function, legal status, and social impact. Recognising administrative detention as a broader category allows for a sharper analysis of the category in general, and reveals patterns in its individual forms. In particular, two aspects of administrative detention stand out.

The first is the dominant role of bureaucratic processes of classification for labelling, dividing, and controlling people into different population categories. Such classification is a central characteristic of all modern societies (Douglas 1966), although Hirschman observed how the British colonisers adopted this practice particularly enthusiastically (Hirschman 1986). Some categories of people were free from government constraint over these aspects of their lives, while for other categories the control was overbearing. Classification into social sub-groups impacted any number of aspects of people's lives: where they could work, who they could marry, whether they could travel, if they could migrate. Administrative detention is deployed when governments identify certain sub-groups as a threat to national identity or social cohesion. Time and again, the systems of classification used for identifying these sub-groups have been arbitrary, opaque, overly mechanistic, and non-reviewable. For individuals and groups thus classified into a category whose confinement is required by law, these decisions could be devastating.

Second, administrative detention is most often implemented with unmitigated executive control, characterised by a lack of transparency, accountability, and review. Administrative detention creates the conditions for government to impose its authority over different cohorts within the population, out of sight from the rest of the population, and without the hindrance of external oversight. Executive control characterises the whole of the detention experience: from the non-reviewable and automatic way that a person is classified into a detainable category, to the conditions of the detention environment and the day-to-day treatment meted out by staff. In some instances, as in quarantine, the effect was benign. Yet regardless of the outcome of their time in detention, people subject to detention are left in no doubt about the few rights afforded to them, and the tenuous degree to which their presence was tolerated by the nation state.

These same two characteristics of classification and executive control form the foundation of the policy and practice of the contemporary system of immigration detention, to which we now turn. Australia's system of immigration detention

In 1989, a boat of twenty-six asylum seekers from Cambodia arrived on the northern coast of Western Australia. They disembarked on a beach, and met some people who alerted the authorities. They were flown 3000 km to Sydney where they were housed in a migrant hostel for a short time, before being flown back the 3000 km to Port Hedland in Western Australia. The government acquired old dormitory accommodation previously used for miners. They wrapped it in a fence, and thus established Australia's first immigration detention centre. There, they were detained for the remaining time it took for their applications for asylum to be processed: an average of 523 days (Port Hedland detention centre, National Museum of Australia 1991).

Port Headland might be one of Australia's most remote locations. On the map, the town is the furthest point on the mainland from the major metropolitan centres of Melbourne and Sydney. Broome, the closest town, is over 600 km away (Phillips & Spinks 2013). As though its architects were trying to emphasise that they had gone as far as they could, the detention centre sat right on the coastline. Detainees could look through the fencing to the ocean over which they'd travelled.

Three years later, in 1992, when these same detainees made a successful claim in the courts against their unlawful incarceration, Australia's policy of mandatory, indefinite and unreviewable immigration detention was legislated, applicable retrospectively (Lester 2018). In the thirty years since, this system of non-judicial incarceration has been recognised as one of the world's most punitive and damaging.

Throughout the 1990s and early 2000s, the government began to build other detention centres in equally remote locations. The Woomera detention centre, in the red South Australian desert, was a four-hour drive from Adelaide. When detainees looked out from the razor-wire, they would see nothing but the vast expanses of red soil, dotted with scrubby desert bushes. Baxter detention centre was equally as remote, but at Baxter, the facility was built so that people could not look out. Then there was Christmas Island; Curtin, in far north Western Australia; Scherger on the Cape York Peninsula; Northam and Leonora in Western Australia's interior. The common feature is that they are all so very remote, making visits by family, supporters, or lawyers expensive, time-consuming, and logistically difficult. All are on former defence land, so media access and publishing images is not permitted.

Then, during the late 1990s unrest in Afghanistan, Pakistan and other countries meant that more people travelled to Australia by boat, and the immigration detention centres were full. The Howard government made a stand against a Norwegian freighter ship, the MV Tampa, that had rescued 433 mostly Afghans from their sinking Indonesian fishing vessel. It called around our Pacific neighbours, to see if anyone was inclined to help. Nauru, and later Papua New Guinea, agreed to help in exchange for huge increase in development investment and aid. And so, in 2001, at the same time that terrorists attacked the twin towers in New York City, the policy of offshore processing was introduced. All people who arrived to Australia by boat would be sent to Nauru or Papua New Guinea for processing.

Nauru and Papua New Guinea were Australia's only former colonies. Both are very poor, with histories of corruption, underdevelopment, and exploitation. Both have been long dependent on Australia's financial assistance, so when the offer came for more, political leaders saw that it was in their country's economic interests to accept.

Nauru is situated 4000 kilometres from Australia, with a flight time of five and a half hours over the South Pacific Ocean. After Vatican City and Monaco, it is the world's third smallest country, measuring only 21 square kilometres (8 square miles). It is an island made entirely of phosphate, and in the past this has made it very very rich. In the mid-1970s, its GDP per capita measured only second to Saudi Arabia. By the 1990s, the phosphate deposits were exhausted, and mining had depleted 80% of Nauru's landmass, rendering it uninhabitable and unable to be rehabilitated. During the 1990s Nauru's main source of income came from its status as a tax haven, selling passports, and money laundering, but in 2001, when Australia called, it was nearly bankrupt. In Papua New Guinea, the Manus Province is the most remote area of PNG, and desperately underdeveloped. While the capital Port Moresby sits to the south of the country, facing Australia, Manus is to the north, facing the Philippines and beyond that, the Asian powerhouses. This geography means that Manusian residents have long felt isolated and neglected by their capital. They feel exploited too, because Lobrum Island has hosted an army base used by Australia and other allied forces, but the community surrounding it has not benefited from foreign income, and remains severely underdeveloped. It is within this army base that the detention centre was located. Around 1,600 people, mostly from Afghanistan and Iraq, were detained on Nauru or Manus Island between 2001 and 2006. This included 125 women, 213 children and 30 unaccompanied minors on Nauru, and 65 women and 125 children on Manus Island. In addition, 19 babies were born on Nauru, and three on Manus Island. The majority were detained for three years or more. Approximately 70% were eventually resettled in Australia and other countries.

Following the September 11 terrorist attacks, US-led military operations forced regime change in Afghanistan and Iraq, and in 2002 the boats stopped arriving. The policy was eventually abandoned in 2008, but almost immediately the boats began arriving again, carrying larger numbers of people seeking asylum than Australia had seen before. Under political pressure the government, this time led by Labor's Julia Gillard, reopened the centres on Nauru and Manus Island. This time, the policy was that no one who travelled to Australia by boat would ever be let into the country. 4,183 men, women and children were subject to this second iteration, known as Offshore Processing (Refugee Council of Australia 2022).

Meanwhile, the network of detention centres on the mainland remains. Asylum seekers are the minority in this population. Most people detained onshore are people who have breached their visa conditions, usually because they have committed a crime, and are awaiting deportation. Because many of these people have spent a large proportion of their life in Australia, their country of citizenship does not want to accept a deported criminal who is essentially Australian. Some spend many years in detention while their deportation is negotiated. The average length of time someone spends in Australia's immigration detention centres is 689 days, compared with 55 days in the US, and 14 days in Canada (Dinham & Arora 2022).

Executive control over immigration detention

Like other forms of administrative detention, immigration detention is designed and implemented in a way to ensure executive control. For immigration detention, the law has been designed so that a court cannot interfere with the decision to detain, and in fact no one actually makes the decision to detain. Instead, people are detained automatically when they have been assessed to belong within a particular administrative category. No consideration is granted to the age, health, prior life experiences, or character, or other considerations regarding the proportionality of detention.

Executive control is assisted by the high levels of secrecy, which is a designed-in feature of the policy. The first way secrecy is achieved is by controlling media coverage. It has become commonplace for our political leaders to rebuff any questions, stating they do not discuss 'on-water matters', a phrase coined by then Minister for Immigration, and later Prime Minister, Scott Morrison (ABC News 2014). Stymying media coverage also extends to pursuing journalists who publish stories about immigration detention. The Australian Federal Police investigated eight Australian journalists for 'unauthorised disclosure of Commonwealth information', including those who had written about child sexual abuse on Nauru.

Since 2001, governments have constructed the arrival of people seeking asylum by boat as a national security crisis, justifying a military response and further non-disclosure. As Former Prime Minister Tony Abbott explained,

> We are in a fierce contest with these people smugglers. And if we were at war, we wouldn't be giving out information that is of use to the enemy just because we might have an idle curiosity about it ourselves (Burnside 2015).

Secrecy is also built into the law. The Australian Border Force Act (2015) prohibits anyone working under the Act (either employed directly by the government or contracted to it) from speaking about their work to anyone, not the media, not to other organisations, not even to their partners. Those that breach the Act risk two years imprisonment. The terms of the Act were so broad that, as human rights lawyer Julian Burnside quipped, it makes it an offence to report an offence (Burnside 2015). After two years of concerted lobbying, the Act was amended to allow medical professionals to speak to each other about patients receiving their treatment (Booth 2015).

Finally, a 'pervasive culture of secrecy' dominates the relationship between Australia and the private corporations that run the centres (Committee of Australian Senate 2015). We are not privy to the reporting requirements contained within the commercial-in-confidence contracts, but we can ascertain that they contain disincentives for the companies to report any incidents or concerns about the well-being of people detained.

In its activity against journalists, Australia was assisted by Nauru. In 2014, Nauru raised the amount it costs to lodge an application for a journalistic visa to the country from A$ 200 to 8000, non-refundable even if the visa is refused, as they all are (Davidson 2018a). In 2016, in response to a journalist gaining entry on a tourist visa, Nauru announced that it would refuse entry to all visa requests from Australian and New Zealand passport-holders unless they were contracted by the Australian government (Davidson 2016). When Nauru hosted the important Asia-Pacific Economic Forum in 2018, it denied entry to all Australian journalists, and detained a highly esteemed New Zealand journalist for speaking with refugees during her visit (Davidson 2018*b*).

Life in offshore processing

Despite all these efforts, however, we know a lot about what life was like within the offshore processing centres on Nauru and Papua New Guinea. We know because refugees managed to communicate their experience, helped by collaborators and contraband mobile phones, and their accounts have been supported by whistle-blowers. For example, Abdul Aziz Muhamat used a smuggled mobile phone to exchange WhatsApp voice memos several times daily with his collaborator, Michael Green, based in Melbourne. Abdul tells Michael of his family and life in Sudan, how he came to be detained on Manus Island, and his life there. As he does, we hear the soundtrack of life in detention: the deep inhalation of cigarette smoke, the crunching of his footsteps on the gravel as he paces, and the slow-moving despair of endless boredom, so endless it is a violence in and of itself. The 3500 memos were compiled into *The Messenger* podcast, which won international human rights awards (Meade 2017).

We know of the violence of the centres through the writing of Behrouz Boochani, who wrote his powerful novel *No Friend but the Mountains* in text messages to his translator Omid Tofighian (Boochani 2018). Boochani details the antagonistic relationship between detainees, guards and the local residents, who resent the detention centre and do not see any benefits for their community, especially not the massive profits. The antagonism between locals and detainees flares up periodically, as it did most seriously in 2014 when locals and guards broke into the centre and rioted. More than 70 detainees were injured, and one detainee, Reza Berati, was killed (Doherty 2021).

From Nauru, we have photographs of tents with toxic mould so seemingly alive that it hangs in tendrils from the top of the canvas tents, fed by the unending heat and humidity. Former staff have sued the Australian government for the so-called 'black mould illness' they developed, leaving them with cognitive impairment and chronic lung infections. A microbiologist contracted to assess the risk blew the whistle after his report was covered-up (Doherty 2018). We have photographs of toilet facilities blocked and overflowing. Running water is restricted to a few hours each day, and is not potable. People may wash their clothes once a fortnight, or if water is limited, once a month. People were required to queue for hours, without shade, for two Panadol tablets, a daily ration of women's sanitary items, or to get their fingernails cut by the nurse.

We also know about the violence, abuse and neglect from the people who once implemented the system. A former head of the Australian Border Force, Roman Quadvelieg, said that in that role he thwarted and obstructed applications for medical repatriation. In 2017-2018, the government fought 34 applications for transfer to Australia for medical treatment in the Federal Court; including many with children as the primary applicant. In 81 other cases, applicants were transferred only after a lawyer was engaged. These cases included that of a woman who requested transfer to Australia for an abortion after falling pregnant from a rape. She was eventually sent to PNG for the abortion, despite it being illegal and unsafe there. Several of the people who have died in offshore processing have done so from treatable conditions.

The situation for children on Nauru was particularly disturbing. In 2014, nine Save the Children employees were instantly removed from Nauru when they reported child sexual abuse, accused of 'political activism' (White 2017). Two government inquiries upheld their accounts. In 2016, the Guardian published a leaked cache of over 2000 incident reports it called the 'Nauru Files'. Over half of these files related to incidents involving children, including assaults, sexual abuse, child abuse and self-harm attempts (Farrell et al. 2016).

More than 200 children were detained on Nauru since 2013, including 46 babies born there. The last were released in 2019. During the days they attended school, where some signed their drawings and schoolwork with their boat numbers. At the end of the day they returned to the camp, where there are no toys or play equipment. While the physical and mental health of their parents and other adult carers declined, most reported they had witnessed violence, sexual abuse, and self-harm. Children as young as eight have attempted suicide, and there were countless incidents of self-harm. Thirty children suffered a rare condition which doctors call 'resignation syndrome' or 'traumatic withdrawal syndrome', in which they stop talking, eating, toileting, and effectively put themselves into a coma (Newman 2018).

Isolation of asylum seekers: immigration detention in Australia

It is difficult to ascertain the purpose of this situation. In my view, the most satisfying explanation is the argument that people are treated so poorly so that they are convinced to abandon their claims to refugee status and return to their country of origin. The former head of the International Health and Medical Service, Peter Young, argued that as such, offshore processing fulfils the definition of torture:

> If we take the definition of torture to be the deliberate harming of people in order to coerce them into a desired outcome, I think it does fulfil that definition... This detention is created in such a way as to act as a deterrent, to encourage people to return [to their homeland], and to stop other people trying to seek asylum. The harmfulness is a 'designed-in' feature. You can't allow transparency, if what you're trying to do is inflict suffering. Secrecy is necessary because these places are designed to damage (Metherell 2014).

Or, as Kurdish journalist and former detainee Behrouz Boochani, put it, 'the pain is there to send you home' (Boochani 2016).

In July 2022, 112 people remain in Nauru, and 105 in Papua New Guinea. All live in the community, and have some degree of freedom. Because the policy remains that they will never be allowed to enter Australia, there are few options for them. Australia signed a deal with Cambodia, which is still open, that it would accept refugees who chose to go to that country voluntarily. It was a good deal for Cambodia, which was worth $ 55m, because only 7 asylum seekers chose to travel there, and of them, only one stayed longer than a year. A more significant deal was struck with the US when Barack Obama was president, and small numbers of detainees have been selected for resettlement. The deal was upheld by President Trump, although he called it a 'dumb deal' (Murphy & Doherty 2017), but his list of states banned from immigration meant that many of the refugees were ineligible during his Presidency. Since 2013, New Zealand has had a standing offer to resettle refugees from Nauru and Manus Island, which Australia accepted only in 2022 (Grattan 2022).

In 2022, after nine years detention, Mehdi Ali received a resettlement place in the US. He had spent more than a third of his life in Australia's detention system. He now lives in Minnesota. He describes the experience of detention, as leaving a 'permanent scar'. Of his new home he has said, "It's good to go out of your place without pat search or handcuffs... Cook a good meal, go to restaurants. It's good to enjoy the live music and take a walk in nature (Osborne 2022)."

Amy Nethery

Conclusions

The convict settlement of Australia was established at the same time that English prison reformers were advocating for a less brutal approach to prisons. Apart from a short-lived experiment on Norfolk Island, authorities in Australia did not heed these calls for reform. Incarceration in all its forms proved too valuable a tool for establishing control over the new colony. Administrative detention in particular, free of the pretence of legal objectivity, allowed governments to manage internal and external boundaries, and incarcerate and control whole groups of people. Thereafter, this form of incarceration became a template to which future policymakers have reached time after time to manage perceived threats to national identity, integrity, or security.

In all such instances, immigration detention is presented as a core function of Australian sovereignty and thus exempt from democratic checks-and-balances, pluralist policymaking, deliberative public debate, or constraint by liberal international institutions. It is notable that the Australian polity – either as British subjects or later, Australian citizens – have never sought to temper the power of the executive arm of government on these questions of membership. Unlike other settler colonial states, Australia does not have a bill of rights or a treaty with its indigenous peoples. Neither have administrative detention policies met sustained resistance in parliaments: since Federation, both major parties have generally offered unconditional support for administrative detention policies, which are viewed as the domain of the government. The judiciary has largely regarded decisions of inclusion and exclusion to be the purview of government, and both of Australia's major parties have collaborated to incrementally restrict judicial oversight (Crock, 2004).

Neither have international instruments mitigated executive control. Over the last decade Australia has effectively de-ratified the Refugee Convention from its domestic laws. Australia has also strongly resisted the Optional Protocol to the Convention against Torture and Other Cruel, Inhuman or Degrading Treatment or Punishment (OPCAT) which would allow inspection without notice of all sites of incarceration. It has refused to sign the Global Compact on Safe, Orderly and Regular Migration because it would require Australia to abandon its commitment to immigration detention.

This essay has presented an endogenous history of Australia's attachment to administrative detention. But no idea develops in true isolation. Australia has also been influenced by outside ideas: the US's detention of Haitian refugees in Cuba's Guantanamo Bay in the 1980s formed the model for its offshore detention centres, for example. In more recent years, it is Australia's approach that has become the model for some countries seeking new ways to manage large numbers of people seeking asylum. The policy, as I have shown, is deeply harmful and fundamentally undemocratic, and should not be replicated.

Amy Nethery

References

ABC News (2014), 'Prime minister Tony Abbott likens campaign against people smugglers to "war"', http://www.abc.net.au/news/2014-01-10/ abbott-likens-campaign-against-people-smugglers-to-war/5193546.

Boochani, B. (2016), 'In Manus prison, the pain is there to send you home *The Saturday Paper*', https://www.thesaturdaypaper.com.au/news/politics/2016/04/16/manus-islands-appalling-health-care-record/14607288003132#hrd.

Boochani, B. (2018), *No friend but the mountains: Writing from Manus prison*, Pan Macmillan, translated by O. Tofighian, Sydney, Australia.

Booth, A. (2015), 'Health workers exempt from immigration detention secrecy provisions, *SBS news*', https://www.sbs.com.au/news/article/health-workers-exempt-from-immigration-detention-secrecy-provisions/a02f3c1ri.

Bosworth, M. (2014), *Inside Immigration Detention*, Oxford University Press, Oxford, U.K.

Burnside, J. (2015), 'Why can't we know what's happening on Nauru and Manus island? *The Guardian*, 8th of October', https://www.theguardian.com/commentisfree/2015/oct/08/why-cant-we-know-whats-happening-on-nauru-and-manus-island.

Committee of Australian Senate (2015), 'Taking responsibility: Conditions and circumstances at Australia's Regional Processing Centre in Nauru (select committee on the recent allegations relating to the conditions and circumstances at the Regional Processing Centre in Nauru)', https://apo.org.au/sites/default/files/resource-files/2015-09/apo-nid56996.pdf.

Davidson, H. (2016), 'Nauru bans entry for Australians and New Zealanders without a visa (*The Guardian*, 19th February)', https://www.theguardian.com/world/2016/feb/19/nauru-bans-entry-australians-new-zealanders-without-visa.

Davidson, H. (2018*a*), 'Australia jointly responsible for Nauru's draconian media policy, documents reveal (*Guardian*, 4th October)', https://www.theguardian.com/australia-news/2018/oct/04/australia-jointly-responsible-for-naurus-draconian-media-policy-documents-reveal.

Davidson, H. (2018*b*), 'New Zealand reporter detained by police on Nauru after refugee interviews', https://www.theguardian.com/world/2018/sep/04/new-zealand-reporter-detained-on-nauru-after-refugee-interviews.

Dinham, A. & Arora, A. (2022), 'Inhuman: Legal, human rights experts slam laws allowing detention in Australia', https://www.sbs.com.au/news/article/inhuman-legal-human-rights-experts-slam-laws-allowing-detention-in-australia/g3imozulz.

Doherty, B. (2018), 'Nauru mould problem was of "epic proportions", microbiologist says, *The Guardian*, 24 February', https://www.theguardian.com/world/2018/feb/23/nauru-mould-problem-was-of-epic-proportions-microbiologist-says.

Doherty, B. (2021), 'Slow transfer to Australian hospital contributed to death of Iranian refugee on Nauru, coroner finds', *The Guardian* (https://www.theguardian.com/australia-news/2021/nov/01/slow-transfer-to-australian-hospital-contributed-to-death-of-iranian-refugee-on-nauru-coroner-finds).

Douglas, M. (1966), *Purity and Danger: An Analysis of the Concept of Pollution and Taboo*, Routledge, London, U.K.

Farrell, P., Evershed, N. & Davidson, H. (2016), 'The Nauru files: cache of 2,000 leaked reports reveal scale of abuse of children in Australian offshore detention, *Guardian*, 10 August', https://www.theguardian.com/australia-news/2016/aug/10/the-nauru-files-2000-leaked-reports-reveal-scale-of-abuse-of-children-in-australian-offshore-detention.

Fassin, D. (2017), *Prison Worlds: An Ethnography of the Carceral Condition*, John Wiley & Sons, London, U.K.

Foucault, M. (1967), *Madness and Civilization: A History of Insanity in the Age of Reason*, Routledge, London, U.K.

Foucault, M. (1977), *Discipline and Punish: The Birth of the Prison* (translated by A. Sheridan), Vintage Books, New York, U.S.A.

Goffman, E. (1961), *Asylums: Essays on the social situation of mental patients and other inmates*, Aldine Transaction, New Jersey, U.S.A.

Grattan, M. (2022), 'Morrison government finally accepts deal with New Zealand to resettle refugees *The Conversation*, 24 March', https://theconversation.com/morrison-government-finally-accepts-deal-with-new-zealand-to-resettle-refugees-179949.

Hirschman, C. (1986), 'The making of race in colonial Malaya: Political economy and racial ideology', *Sociological Forum* **1**, 330–361.

Human Rights Watch (2016), 'Australia: Appalling abuse, neglect of refugees on Nauru', https://www.hrw.org/news/2016/08/02/australia-appalling-abuse-neglect-refugees-nauru.

Kaldor Centre (2021), 'Medical transfers from offshore processing to Australia', https://www.kaldorcentre.unsw.edu.au/publication/medevac-law-medical-transfers-offshore-detention-australia.

Lester, E. (2018), *Making Immigration Law: The Foreigner, Sovereignty, and the Case of Australia*, Cambridge University Press, Cambridge, U.K.

Meade, A. (2017), 'The Messenger podcast wins top honour at international radio awards, *The Guardian*, 20 June.', https://www.theguardian.com/australia-news/2017/jun/20/the-messenger-podcast-wins-top-honour-at-international-radio-awards.

Mehdi, A. (2022), 'A letter to Australia's immigration minister from a refugee', *Al Jazeera* (https://www.aljazeera.com/features/2022/2/13/a-letter-to-australias-immigration-minister-from-a-refugee).

Metherell, L. (2014), 'Immigration detention psychiatrist Dr Peter Young says treatment of asylum seekers akin to torture, *ABC News*', https://www.abc.net.au/news/2014-08-05/psychiatrist-says-treatment-of-asylum-seekers-akin-to-torture/5650992.

Monash University (2022), 'Australian border deaths database', https://www.monash.edu/arts/migration-and-inclusion/research/research-themes/migration-border-policy/australian-border-deaths-database.

Murphy, K. & Doherty, B. (2017), 'Trump lashes 'dumb deal' with Australia on refugees after fraught call with Turnbull *The Guardian*, 2 February', https://www.theguardian.com/australia-news/2017/feb/02/trump-told-turnbull-refugee-agreement-was-the-worst-deal-ever-report.

Nagata, Y. (1996), *Unwanted Aliens: Japanese Internment in Austalia*, University of Queensland Press, Queensland, Australia.

Nethery, A. (2021), 'Incarceration, classification, and control: Administrative detention in settler colonial Australia', *Political Geography* **89**, 102457.

Nethery, A. & Holman, R. (2016), 'Secrecy and human rights abuse in Australia's immigration detention centres', *International Journal of Human Rights* **20**, 1018–1039.

Newman, L. (2018), 'Explainer: What is resignation syndrome and why is it affecting refugee children? *The Conversation*', https://theconversation.com/explainer-what-is-resignation-syndrome-and-why-is-it-affecting-refugee-children-101670.

Osborne, J. (2022), '*I could not forgive*: Mehdi Ali on the scars from nine years of immigration detention', https://www.sbs.com.au/news/insight/article/i-could-not-forgive-mehdi-ali-on-the-scars-from-nine-years-of-immigration-detention/g9l20fjqn.

Parks Australia (2001), 'Millions of crabs march on christmas island', https://parksaustralia.gov.au/christmas/news/millions-of-crabs-march-on-christmas-island/.

Phillips, J. & Spinks, H. (2013), 'Immigration detention in Australia: Background note, Australian Government Parliamentary Library', https://www.aph.gov.au/about_parliament/parliamentary_departments/parliamentary_library/pubs/bn/2012-2013/detention.

Port Hedland detention centre, National Museum of Australia (1991), https://www.nma.gov.au/defining-moments/resources/port-hedland-detention-centre.

Refugee Council of Australia (2022), 'Offshore processing statistics', https://www.refugeecouncil.org.au/operation-sovereign-borders-offshore-detention-statistics/.

Reynolds, H. (1989), *Dispossession: Black Australians and White Invaders*, Allen and Unwin, Sydney, Australia.

Rothman, D. J. (1971), *The Discovery of the Asylum: Social Order and Disorder in the New Republic*, Aldine Transaction, New Brunswick U.S.A.

Saunders, K. (1993), *Minorities in wartime*, Bloomsbury Academic, London, U.K., chapter 'Inspired by patriotic hysteria?' Internment policy towards enemy aliens in Australia during the second world war.

Thwaites, R. (2014), *The Liberty of Non-citizens: Indefinite Detention in Commonwealth Countries*, Bloomsbury Publishing, London, U.K.

White, S. (2017), 'Government pays compensation to Save the Children workers removed from nauru, ABC News', https://www.abc.net.au/news/2017-01-31/save-the-children-workers-government-pays-compensation/8217686.

Wilson, R. (1997), 'Bringing them home: National inquiry into the separation of Aboriginal and Torres Strait islander children from their families', Commonwealth of Australia, Human Rights and Equal Opportunity Commission, report, https://humanrights.gov.au/sites/default/files/content/pdf/social_justice/bringing_them_home_report.pdf.

3 The closeting of secrets

ADRIAN KENT
University of Cambridge

Abstract: The definition and properties of information may seem to be fundamental features of the world that are independent of how particles, fields and space-time behave. In fact, though, information is fundamentally physical and twentieth century physics has radically changed our understanding of its nature and properties. Einstein's relativity theories tell us that information cannot travel faster than the speed of light in vacuum. Quantum theory tells us that the information carried by microscopic systems is qualitatively different from the familiar "classical" information with which we presently communicate and compute: for example, quantum information cannot be copied. These realisations have led to new applications and emerging new technologies, including relativistic quantum cryptography and new forms of quantum communication and computation in space-time. This lecture will illustrate several ways in which physics-based cryptography and communication allow otherwise unachievable forms of security and flexibility, including guaranteeing a fair coin toss for mistrustful parties, making and later revealing secret predictions that carry a guaranteed time stamp, and secure forms of money that emulate quantum particles by following multiple paths and recombining to solve otherwise insoluble trading problems. We will also ask how confident we should be that we now fully understand how information is carried and processed in nature, and whether new physics discoveries might yet change our understanding and lead to further technological advances.

Introduction

The insight that information is physical (spelled out, for example, by Landauer 1991) gave us a new way of thinking about fundamental physics and a new wave of information technologies. In a sense, the physicality of information is obvious: computing, communication and cryptography use physical systems and follow physical laws. To send or receive information also requires the transport of physical entities such as a radio wave, a book, a pulse of light, an elementary particle, or a gravitational wave. Computers are physical devices that are set in some initial state, to follow some algorithm, which ultimately is determined by the laws of physics, to reach a final outcome. Secure messages between a pair of computers requires them to be linked by a communication channel that carries physical information. More generally, any cryptographic task ultimately involves controlling access to information via a physical mechanism that inspires confidence that the information is not available to the wrong people.

This would have been an academic point if all forms of physical information were essentially equivalent, as was implicitly assumed through most of last century in computer science, communication theory and cryptology. Starting with the theoretical work of Turing (1936) and the early designs of computer architecture by von Neumann (1945), computers were modelled as devices that have a finite number of registers, each in one of a finite number of possible discrete states, with those states changing at discrete times following deterministic laws. Shannon and others (e.g., Shannon 1948, Shannon & Weaver 1949) developed the mathematical theory of information as a quantity that can be represented by a finite string of characters drawn from a finite alphabet. These are powerful models. It remains the case today that they do characterise all the types of computer and forms of communication that are in widespread use. But everything in these models – paper tapes on which symbols can be written and from which they can be read, devices that move the tape around depending on the symbols, discrete pulses of radio waves, and so on – can be described by *classical* physics, using Newton's or Maxwell's equations. As we now understand it, these models characterise *classical* information, and give us classical theories of computation, communication and cryptography.

Quantum physics tells us that elementary particles, atoms and ions, weak electromagnetic pulses and other microscopic physical systems behave qualitatively differently from systems described by classical physics. It took decades to fully appreciate that their informational aspects are most naturally described not by combining classical models of information with new physical constraints, but in terms of a new, radically different but equally coherent model, quantum information. Its properties can be exploited to produce distinctive and in some important respects more powerful computational algorithms, new and more secure ways of controlling access to information, and qualitatively new forms of communication – hence the development of quantum computing, quantum cryptography and quantum communication as theories and emerging new technologies.

Before quantum theory was fully developed, twentieth century physics had already seen two other revolutions, Einstein's theories of special and general relativity, that transformed our understanding of space, time and gravity. However, they have subtler implications for the physics of information than quantum theory does. There are some interesting nuances which may ultimately be important. General relativity predicts black holes, which suggests that information may effectively be lost. It also allows closed time-like curves, which require information to flow in a self-consistent causal loop. Still, at least for present-day practical purposes, special and general relativity, like classical Newtonian mechanics, and unlike quantum theory, tell us that physical systems, can be measured as precisely as we wish; that the outcomes of measurements on a system are determined completely by its state; and that the state always follows deterministic evolution laws. In other words, physical systems in special and general relativity carry only classical information.

Special and general relativity do imply a new constraint on information, that it cannot be transmitted faster than light can. This gives a physically guaranteed way to control the flow of information: if secret information is broadcast at a particular point in space and time, it will remain secret at any space-time point that a light signal cannot reach from the broadcast point. Such secrecy will be temporary because information broadcast from any given location can eventually reach every point in space. Nonetheless, as we will see, this temporary guarantee turns out to be powerful enough to secure some important cryptographic tasks.

Relativity and mistrustful cryptography

A familiar application of cryptography is in encrypting messages to keep them secure from intermediaries or eavesdroppers while allowing the intended recipient to decode them. Consider Alice and Bob, who wish to communicate while keeping their messages secret from the evil eavesdropper, Eve. The relationships here are straightforward: Alice and Bob trust one another and collaborate; Eve is an untrustworthy adversary. Alice and Bob would ideally prevent Eve's involvement completely but they cannot, so resort to cryptography to achieve the same effect.

Relationships are often more complex, though. Many cryptographic tasks involve two or more parties who must collaborate to generate, process and control information with specified constraints, but who do not necessarily trust one another. Typically, these applications could be securely implemented if there were an agent or device that everyone involved trusts to generate and handle information according to precise stipulations, such as keeping information perfectly secure during a protocol and forgetting and destroying some or all of it afterwards. The parties must thus completely trust the agent's competence and the security of their technology as well as their honesty – strong requirements that may not be satisfiable in the real world.[1]

Mistrustful cryptography aims to design schemes with security guarantees that eliminate the need for trust. In the standard model of mistrustful cryptography, each party trusts no one else. More precisely, they trust the security of a laboratory in which they work and that devices inside the laboratory function as specified – we may assume they built and maintain the laboratory and the devices – but nothing outside. It follows that different parties' secure laboratories should not overlap physically: if they did, the overlap region would be trusted to be secure by two or more parties. Each party assumes that everything in the world outside their secure laboratory is potentially under the control of adversaries. There is thus no need to introduce Eve in this model, even for two-party protocols: for Alice, anything that any third party might potentially do to evade the intended constraints on the information in the protocol could equally well be done by Bob, and vice versa.

One of the simplest examples, introduced by Manuel Blum (1983), is secure remote coin flipping. In Blum's scenario, Alice and Bob are separated, have an authenticated communication channel, and wish to generate a string of random numbers, as though they were flipping a mutually trusted fair coin. They may, for example, have a common interest in testing a hypothesis by a method that involves many random choices. The result may have different implications for Alice and Bob, and neither of them knows which result the other would prefer. So, they both want to be sure that the test is reliably random. Their problem is that neither of them trusts the other not to bias the outcome, and they also don't know which way the other may be motivated to bias. Each has their own random number generators – figuratively, their own fair coins – which they do trust. So, Alice could phone Bob, suggest they use her fair coin, and announce the result a. Let us call coin flip results 0 or 1 rather than "heads" and "tails". Alice thus announces $a = 0$ or $a = 1$. But Bob won't trust this as a random outcome. Even if Alice flips the coin on a video link he will think she may be faking it, perhaps

---

[1] Who could you use as a trusted third party to authenticate your bank card at an ATM, and how would the authentication process work? Who could the US and Chinese governments trust to handle, process and destroy sensitive information according to a given protocol?

Adrian Kent

having cut seamlessly to a pre-prepared video chosen to give the result she wants. Bob can suggest they use his coin instead, but of course now Alice has the same worries.

There are possible partially secure solutions – for example, Alice sends her result in code, Bob guesses, and Alice then decodes her message for Bob, with the protocol that the outcome is 0 if Bob guesses right and 1 otherwise. But this needs Alice to be confident that Bob can't break her code during the protocol and Bob to be confident that Alice's message can only be decoded one way. They would prefer an unconditionally secure scheme, that doesn't rely on any assumptions about codes being difficult to break.

The key idea here, and the basis of all relativistic cryptographic schemes (Kent 1999), is to consider Alice and Bob not as individuals but as two networks of collaborating agents – perhaps working for two companies or governments – who can be dispersed at different locations. This allows a simple solution whose security is guaranteed by the light speed bound on information transmission, as illustrated in Figure 3.1. Alice's agent A_0 occupies a secure site at or very near to some location x_0, while Bob's agent B_1 occupies a secure site at or very near to x_1, where the locations x_0, x_1 are pre-agreed, with some agreed small positioning errors allowed. Bob's second agent B_0 occupies another secure site at some location close to x_0, and Alice's second agent is similarly located close to x_1, where "close" means "small compared to the distance $d(x_0, x_1)$". Then, at some agreed time t (with small pre-agreed timing errors allowed), A_0 sends her coin flip outcome a to B_0, and B_1 sends his coin flip outcome b to A_1. The agents all compare notes over authenticated channels, and they take the agreed random number to be 0 if $a = b$, and 1 otherwise.[2] The separation between the pairs of agents is relatively large compared to the short distances between the agents in each pair. So, no information can travel from the point where A_0 sent her outcome a to the point where A_1 received the outcome b from B_1. This guarantees to Alice that Bob's bit b must have been generated in ignorance of her bit a. Similarly, Bob is guaranteed that a must have been generated in ignorance of b. It is thus intuitively clear, and easily provable, that neither of them can influence this overall outcome by biasing their own coin flip, because they must report their result before they learn the other party's. However they choose to generate their own bit, the other party's random number randomises the final outcome.

---

[2] Programmers and logicians will recognise this as the "exclusive or operation", $a \oplus b$, that is true if and only if its arguments differ.

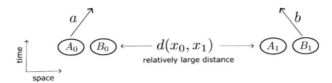

Figure 3.1 An unconditionally secure relativistic protocol for mistrustful coin tossing. Alice's agents A_0 and A_1 occupy secure laboratories prepared by Alice; B_0 and B_1 occupy nearby secure laboratories prepared by Bob. The laboratories are depicted in 2D space at a particular point in time: in the simplest implementations they remain in the same physical region throughout the protocol. A_0 sends a random bit a generated secretly by Alice to B_0. At nearly the same time, B_1 sends a random bit b generated by Bob to A_1. After communicating the bits received, all agents agree on the outcome $a \oplus b$. Provided at least one party honestly inputs a genuinely random number, the outcome is randomly generated and cannot have been biased by either party.

We should examine the assumptions carefully here. Alice must trust not only that her own random number generator (coin) is genuinely random, but also that Bob can't learn in advance what numbers it will produce; similarly Bob. So, we need to assume that each party's laboratories are secure, that they can't be probed or bugged by the other, and that they contain trustworthy random number generators. This may not always be true, of course, but without some such assumptions essentially every cryptographic scheme is vulnerable. If the parties can't trust that anything anywhere, even their own brain, is secure against others' scanning or influence, then they may be beyond cryptographic help.

The appeal to relativistic signalling constraints involves further assumptions. Alice and Bob need to be sure that the message A_1 receives from B_1 is independent of the message A_0 sends to B_0. So A_0, A_1, B_0 and B_1 all need trusted clocks in their laboratories. They also need to know their own locations, or at least each party's agents need to know their separations, i.e., Alice must know the separation $d(A_0, A_1)$ fairly precisely, and Bob must similarly know $d(B_0, B_1)$. The standard cryptographic framework allows trusted devices, so trusted clocks, per se, are a reasonable assumption. Trusting one's own location may also be reasonable in most scenarios, but it is a further assumption. If one adopts the purest form of cryptographic paranoia, then *nothing* outside the secure laboratories can be trusted. A_0 may have what seems a lovely view of Darwin gardens confirming that her secure laboratory is in Newnham Grange, but the images of trees and passing punts could be being generated by Bob, along with fake GPS signals and other location data, to fool her. If this is a real concern, each party needs to house both their agents in a single secure laboratory, which they can efficiently do by making the laboratory long and thin, with one agent at each end. This lets the agents verify their separations using trusted rulers within their secure laboratories.

Crucially, neither party needs to trust the location of the other party's agents (Kent 2005b). A_1 needs only to check the time at which she received the message from B_1 and compare it to the time at which A_0 sent her message to B_0. If B_1 is further away from A_1 than the protocol allows, he will have to send his message earlier. But this can only make the protocol potentially insecure against Bob: if Alice identifies B_1's true location she may station an agent nearby to intercept the message sooner than Bob expects and send it to A_0. Cryptographic security is defined to protect each party if the other deviates from the protocol, not if they themselves do – the latter is not generally possible, since a party can always give away their own secrets if they choose to.

Another theoretical concern is that light speed signalling constraints might not take their expected form in the region where the protocol takes place. Indeed, general relativity tells us that the possible paths for light, and hence for information, depend on local gravitational fields. Neither Alice nor Bob can measure and trust the gravitational field precisely everywhere that this might be relevant. However, unless the gravitational fields are very strong, any effect is small and easily accommodated by allowing a little room for error in the protocol timings. Admittedly, in principle one party could make the effect large enough to cheat by moving a large black hole or neutron star close to the Earth, but even the most paranoid of cryptographers is unlikely to lose sleep over this concern in the foreseeable future.

No-cloning and quantum money

The possibility of using quantum information for cryptography was first pointed out in a truly pioneering paper by Stephen Wiesner (1983), first submitted and rejected in 1970, and published only in 1983, after Wiesner's ideas had been promoted along with other major new developments (Bennett & Brassard 1985) by two other pioneers of the field, Charles Bennett and Gilles Brassard. Wiesner's key insight was that there is a cryptographically exploitable difference between the amount of information required to describe the state of a quantum system and the amount of information anyone can learn about the state from the system itself. Even the simplest (two-dimensional) quantum systems can be prepared in an infinite number of states, which turn out to correspond to the points of a sphere. So, it needs an infinite amount of information to describe their state precisely. Nonetheless, given such a system in an unknown prepared state, however you choose to measure it, you can learn at most one bit of information – the answer to one binary question.

This has no classical parallel: in principle anyone can learn everything about a classical system by carefully measuring all its dynamical variables. This isn't possible for quantum systems, for two reasons. First, quantum measurement results are generally unpredictable, *even if* you know the state precisely. Second, measurements generally change the state of a system: learning information about one variable of a quantum state by a measurement generally diminishes the information obtainable about other variables.

Suppose that you're given a two-dimensional quantum system in a state corresponding to a randomly chosen point on the surface of a sphere, with no other information. You can get some limited (and probabilistic) information about the state – for example, whether the point was likelier to have been closer to the north or south poles before you measured it. But you can't learn which point on the sphere represents the state. If you could, you could make any number of copies of it. Conversely, if you could make even one perfect copy, you could make an arbitrary number of copies by repeating the operation, and you could then carry out different measurements on each copy. The rules of quantum measurement imply this would give arbitrarily precise information about the state, which quantum theory forbids. So, it can't be possible to perfectly clone an unknown quantum state.

Although a formal statement and proof of this "no-cloning theorem" was published only in 1982 (Wootters & Zurek 1982, Dieks 1982), Wiesner had the correct intuition that cloning and other related hypothetical operations that would effectively "amplify" quantum information are impossible. This was the basis of his idea for "unforgeable subway tokens". These combine public classical information, in the form of a visible serial number, with private quantum information, a series of quantum states created by the issuer and stored in the token. The issuer keeps a secret record of each serial number and the corresponding quantum states, so they can check whether the token is valid by carrying out a suitable measurement on the quantum states. Token users can easily duplicate the serial number, but don't know the quantum states, and so the no-cloning theorem means that they can't forge a copy of a valid token. More generally, it can be proven that they

Adrian Kent

can't gain a financial advantage by any strategy (Molina et al. 2012).

The same principle applies, of course, to anything designed to allow access to any service or resources, so these days we generally refer to Wiesner's invention and later extensions as quantum money (when the token is playing a role similar to cash, as in Figure 3.2), or quantum secure tokens (for applications like authenticating logins to a data network).

We will discuss the practical problems in implementing these beautiful theoretical ideas below. First we consider another cryptographic protocol, bit commitment, which, inter alia, may offer a way around some of these problems.

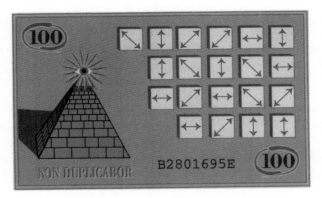

Figure 3.2 An unforgeable quantum money banknote. The serial number is easily copied, and the physical form of the banknote is also copiable, except for the 20 qubits stored in quantum memory – here displayed as their classical descriptions, which are known to the issuer but not the user. The classical or quantum information from these states obtainable by the user is strongly limited by quantum theory. In particular, the user cannot make two or more copies.

No-summoning and bit commitment

Bit commitment (Brassard et al. 1988) is another important two-party mistrustful cryptographic task. Here Alice wishes to give Bob some message in encrypted form, in such a way that she can reveal the unencrypted data for him later, if she wishes. She needs a guarantee that he cannot break the encryption and read the message without her help. He needs a guarantee that she is genuinely committed after the initial exchange, and so cannot, for example, later choose between different decryptions that reveal different data.

A possible but insecure way of doing this is illustrated in Figure 3.3. Since all data can be written in binary, a protocol that commits Alice to a single bit, either 0 or 1, suffices: we can then repeat this protocol bit by bit to commit the full message.

Figure 3.3 An imperfectly secure bit commitment protocol. Alice writes her prediction on paper, puts it in a safe, and gives Bob the locked safe. Later, if she wishes to reveal the prediction, she gives Bob the safe combination. This is insecure against Bob, who may be able to crack the safe or scan its contents. It is also insecure against Alice, who may, for example, put several different predictions on different sheets of paper into a safe that has several different combinations and is designed to destroy all but one sheet when opened, with the surviving sheet depending on the combination used.

Bit commitment is important in its own right. Alice might, for example, want to persuade Bob that she is good at predicting future stock prices, without giving him free predictions. Or she might want to establish priority for a scientific paper or invention, without immediately revealing it to the world. It is also a building block for other cryptographic tasks, including secure voting schemes and auctions. Like coin flipping, it cannot be implemented with unconditional security by standard non-relativistic cryptography, in which Alice and Bob exchange classical and/or quantum information (Mayers 1997, Lo & Chau 1997). However, relativistic cryptography allows various types of protocol for unconditionally secure bit commitment (e.g., Kent 2005a, 2011).

One particularly simple protocol for this uses the so-called no-summoning theorem (Kent 2013), which combines the control of quantum information implied by the no-cloning theorem with the relativistic principle of no faster-than-light-signalling. We define the task of summoning in a relativistic cryptography setting, with two agencies, Alice and Bob, who have agents distributed at various points in space. As usual, Alice's agents collaborate with mutual trust within secure laboratories, and similarly Bob's, but the two agencies keep some information secret from one another, and their laboratories do not overlap. To initiate the task, an agent of Bob's in the vicinity of an agreed point P prepares a random quantum state, keeping secret which state it is, and gives it to a nearby agent of Alice's. Later, at some other point in space-time, an agent of Bob's will "sum-

Adrian Kent

mon" the state, by asking a nearby agent of Alice's to return it. Alice does not know in advance where the summons will be made; her task is to ensure that she can return it wherever it is requested. Bob does not know what Alice does with the state after he hands it over. In the simplest version of the task, illustrated in Figure 3.4, there are just two possible summoning points, Q_0 and Q_1, and it is possible to send information from P to either of the Q_i. The no-summoning theorem states that there is no strategy allowing Alice to guarantee to comply with the summons.

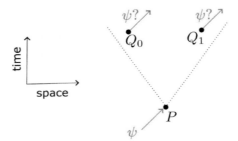

Figure 3.4 A simple impossible summoning task. Alice's agent is given an unknown state ψ at space-time point P. If Bob's local agent requests the state at Q_0, Alice's local agent is required to return the state there; similarly at the space-like separated point Q_1. The dotted lines represent light rays. The no-cloning theorem prevents Alice from making two copies of the state, and the impossibility of faster-than-light signalling prevents her from sending it from Q_0 to Q_1 or vice versa. The no-summoning theorem shows that no other strategy is possible either.

To apply this to bit commitment, let Q_0 and Q_1 be points on two different light ray paths from the point P, where Bob starts the protocol by giving Alice the state ψ. To commit to 0 or 1, Alice sends ψ along a secure channel to Q_0 or Q_1 respectively. If she wishes to reveal her commitment to her chosen bit value b, she returns the state to an agent of Bob's near her chosen Q_b, as illustrated in Figure 3.5. The no-summoning theorem implies that Alice cannot chose the value of b after the point P: in other words, she is genuinely committed. Bob cannot decrypt the committed bit unless Alice chooses to reveal it for the rather Zen-like reason that Alice commits herself without giving him any information – a form of security that makes no sense in conventional cryptography.

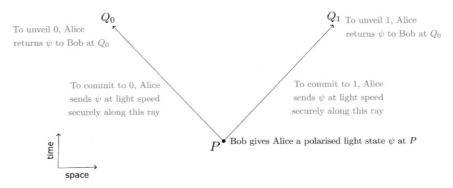

Figure 3.5 A simple relativistic quantum bit commitment protocol based on the no-summoning theorem.

Generalized summoning tasks: when and where can quantum information be in space-time?

There are several mathematically, physically and practically interesting versions of summoning tasks. One nice way of thinking about summoning was introduced by Hayden & May (2016), who pointed out that it gives an operational definition of how, when and where quantum information can be distributed in space-time. As they noted, there is an interesting tension between the properties of quantum information in space, where the no-cloning theorem tells us that quantum states cannot be copied, and its evolution in time. In the approximation where we think of the information as carried by effectively point-like particles, if we keep the particle isolated from interactions then the information stays constant over time. In a sense, it copies itself infinitely often, from each instant to the next. Since special relativity unifies space and time into a single entity, space-time, these seemingly contradictory behaviours must be reconciled in some unified description that respects the symmetries of space-time.

Hayden and May studied this by defining a version of summoning in which there are n pre-agreed pairs of "call points" and "return points", $(c_i, r_i)_{i=1}^{n}$, with each return point r_i in the causal future of the corresponding call point c_i, meaning that information can be sent from c_i to r_i. Bob can "summon" the state at any one of the c_i, and Alice's task is then to return the state at the corresponding r_i, as illustrated in Figure 3.6. They then proved a beautiful result giving necessary and sufficient conditions,

Adrian Kent

depending on simple properties of the geometric relationships between the pairs $(c_i, r_i)_{i=1}^n$, for Alice to be able to guarantee complying with the summoning task

Model for Hayden-May summoning tasks

- State ψ supplied at source points on the network by Bob and Alice. Bob knows ψ but Alice doesn't, so she cannot copy it.

- Both parties have agents throughout the network.

- Both parties know well in advance the list of call-return pairs $(y_i, z_i)_{i=1}^n$.

- At any pre-agreed call point y_i in space-time and agent of Bob's may **summon** the state – requiring Alice to return it at the return point z_i.

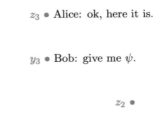

Figure 3.6 Hayden-May summoning tasks. Note that, although we have given an example in 1D space for ease of illustration, the sufficiency conditions become non-trivial only in 2D or higher.

The necessity of the conditions is evident, as it follows directly from the no-summoning theorem and the impossibility of sending signals faster than light. What's fascinating about the result is that their sufficiency is not at all obvious, indeed rather counter-intuitive. To prove it requires clever use of techniques known as quantum secret sharing and quantum teleportation. Quantum secret sharing splits up one piece of quantum information into several parts, in a way that means the information can be recovered from specified subsets. Quantum teleportation separates a quantum state into a combination of classical and quantum information, which must be recombined to recover the original state. Non-trivial Hayden-May summoning tasks in general require both, meaning that the tasks cannot be solved by sending the quantum state along a definite path through space-time. Instead, the state must be delocalised and distributed along many different paths, which collectively allow a response to a summons at any of the call points. The quantum information is effectively constrained so that it can appear within any "causal diamond" – the space-time region in the causal future of a call point and the past of the corresponding return point, although even within these regions it does not generally follow a definite path, as the example in Figure 3.7 illustrates.

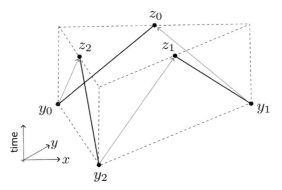

Figure 3.7   A non-trivial Hayden-May summoning task. The black lines, which are the "causal diamonds", are segments of light rays, as are the red arrows. A signal can thus be sent from each call point to either of two return points, but not to the third. Alice can return the quantum state at any one of the z_i, given a call at y_i, but this is not possible if she sends the state on any definite path through space-time. The solution in this example requires quantum secret sharing, splitting the state into three parts and sending one to each y_i. After Hayden & May (2016).

Re-examining the case for quantum money

Although Wiesner described quantum money in the form of physical tokens, its security properties rely entirely on the combination of the classical information in the serial number and the quantum information stored in the token's quantum memory. These aside, the token is just a physical object, with no special security features. Anyone could make any number of "blank" tokens ready for a serial number to be printed and for quantum states to be placed in the quantum memory. So, the physical tokens are dispensable: we can think of quantum money as pure information, with the issuer sending the classical serial number and quantum states to the user via secure authenticated channels.

We've seen that quantum information is unforgeable, but we need to consider carefully precisely what type of security this adds, and in which scenarios it's relevant. It's helpful to compare a purely classical analogue, in which the only security comes from the serial number. To obtain resources, the user simply presents the serial number to an agent of the issuer. In this case, the serial numbers should be sparse – comprising a very small proportion of the numbers of the relevant size – and random, so that it is hard for anyone to guess a valid serial number.

This classical scheme may initially seem to have several drawbacks. One is that it needs penalties (if only the cost in time and effort) to discourage fraudsters from trying repeatedly to guess a valid number until they eventually succeed. But this is also true of quantum money, where the issuer and honest users would also like to discourage fraudsters from repeatedly trying combinations of serial number and randomly chosen quantum states until they are lucky enough to pass the issuer's tests. Another is that the user needs to keep their serial number secure, to prevent fraudsters from copying it and spending the token. Again, though, quantum money requires something similar – the user needs to keep the combination of serial number and quantum information secure. Admittedly, it's arguable that it may be harder for a fraudster to steal quantum information (which they cannot copy) than to copy classical information, and this is a plausible advantage of quantum money. On the other hand, it's also likely easier for malefactors to effectively destroy quantum information, which is fragile and hard to protect (precisely because it cannot be copied) compared to classical information. Also, of course, we currently require users to keep their bank card information and passwords as secure as possible. The associated inconvenience and risks seem to be tolerable, and so one might think that schemes based on classical serial numbers would be similarly viable.

Indeed, bank debit and other electronic cards raise some serious questions about the motivation for quantum money. It may perhaps have seemed plausible in 1970 that unforgeable subway tokens would become a valuable invention, but it's certainly less so now that physical money in general is in rapid decline, most users use electronic cards to access the subway and – crucially – all authentication machines, including toll gates, are easily networked and can quickly exchange data. If a cloned bank card is used for a single metro ride, the loss is small. The fraudulent card used may not be instantly detected, but it will likely be noticed, either by the bank or by the legitimate owner, before large sums are lost. Fraudsters are more likely to try to buy high value items, but (even if this is allowed by a high card limit) these transactions are likelier to initiate cross-checks with the bank and the legitimate user. Legitimate users buying an expensive car on a credit card may thus experience a little delay for cross-checks, but these purchases are not usually so time-critical, and the security justifies any small inconvenience.

One might perhaps think that, in a future world where quantum storage and quantum information transfer are cheap and widely available, the easy transferability of quantum money would be another key advantage. Like present-day cash, but unlike bank debit cards, it could circulate freely, allowing anonymous purchases. But our scheme based on classical serial numbers – which has the same essential functionality as preloaded credit cards, not necessarily tied to a given user – also allows free circulation and anonymous transfer. Both classical serial number tokens and quantum money require either trust, or some form of verification involving the issuer, for third parties to accept them for resources. So, it's not immediately obvious that quantum money has a critical edge here either.

Relativistic signalling constraints motivate quantum money

This still leaves one potentially important application of quantum money: high value transactions that are also highly time critical, in scenarios where relativistic signalling constraints are significant. Imagine a user with agents around the world who may wish to make a high value purchase of shares or commodities, may choose any location, depending on local market conditions, and needs the transaction to go through essentially instantly, so that they are free to trade onwards immediately. Suppose that the quantum money issuer also has agents at all trading locations, all of whom share the information about the serial numbers and quantum states of issued money tokens. The user's local agent, anywhere, can then give the quantum money information (classical and quantum) to the issuer's local agent, who can verify it without cross-checking with other agents.

In our classical alternative scheme, this is insecure without cross-checking. The user could give each of their agents a copy of the serial number and instruct all agents to make a purchase at the same time. If those purchases are instantly accepted, the user succeeds in illegitimate multiple spending of a single money token. To prevent this, the issuer would need to tell each of their agents to check with all the others before accepting a serial number token. If they use radio communications around the Earth, this takes about 0.13 seconds – negligibly short for ordinary transactions, but an unacceptably long delay for some forms of financial market trading. It's true that the issuer, could, alternatively, rely on detecting illegitimate multiple spending after the event. Users planning simple fraud may perhaps be deterred by this and the likelihood of penalties. But one can imagine scenarios in which two or more users apparently independently spend the same token, each claiming that they were the legitimate owner at the point of spending, leaving the issuer with a legally complicated situation, trades that are difficult to unwind, and a loss that may not be easy to recover. Another concern is that agencies such as hostile governments might initiate multiple transactions and onward trades with the aim of large-scale disruption rather than simple profit.

Thinking of the token more generally as something that enables secure access to some resource gives further scenarios where the quantum advantage is important. For example, an agency running a distributed computer network might want roving agents to be able to access one computer as swiftly as possible to defend against cyberattacks, but also to prevent any one agent from accessing many computers simultaneously in case they prove unreliable and implement an attack themselves.

Summoning and quantum money

Given that time is critical in these scenarios, we need to understand just how efficiently quantum money can be propagated to a space-time point that depends on real time data – market prices, events, triggered warnings of an attack on a network – that may be distributed around (or beyond) the Earth. Since the classical serial number data can (at least in principle) be copied and sent at light speed in all directions, it is the unknown and uncopiable quantum information that gives us non-trivial constraints. This is another type of summoning problem (Kent 2018): can we arrange a strategy so that, given incoming data (for simplicity assumed classical) m_i at a list of space-time points P_i, we can guarantee to send the quantum part of the money token to a point $Q(m_1, \ldots, m_M)$ that depends on those data? We assume here that the P_i are known well in advance, since if necessary the list of these points can include every location in space that is being monitored together with every clock tick in time at which data could be received. We can also assume that our traders or network defenders have a predetermined strategy, so that the dependence of Q on the incoming data is known in advance, as illustrated in Figure 3.8.

Hayden-May summoning tasks are a (very) special case of this general summoning problem, in which the points P_i are the call points c_i, the inputs m_i are either 0 (no call at P_i) or 1 (call at P_i), it is guaranteed that

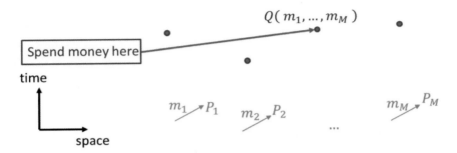

Figure 3.8 A generalized summoning task representing a possible trading strategy. Information arrives at space-time points P_i, implying that the optimal trading point is $Q(m_1, \ldots, m_M)$. The user needs, if possible, to propagate her quantum money state to this point.

precisely one of these, say m_j will be 1, and the point $Q\{m_1, \ldots, m_M\}$ is the corresponding return point r_j. As we've seen, when Hayden-May tasks are solvable, the solution generally requires the quantum information to be delocalized, so that it does not follow a single definite path through space-time. This is also true of most generalized summoning tasks and highlights another key advantage of quantum money over classical physical money tokens. Even if we had effectively unforgeable classical money tokens, and even if we were able to send them at close to light speed, they would necessarily travel along definite paths, and so would not in general be able to solve summoning problems that quantum money can.

A quantum-inspired reconceptualisation of money

Another fruitful way of thinking about this comes from appreciating that the concept of money is not set in stone. Ultimately, money is a tool to solve problems, and the properties we require of it depend on the precise set of problems we need to solve. Special relativity tells us there are properties we would like money to have – being flexible enough to be able to solve summoning problems, while being secure enough to prevent illegitimate multiple spending – that familiar forms of money lack. This suggests a reconceptualization of money, as something that has these properties, perhaps among others. Quantum theory tells us that it is indeed physically possible to have these properties, and so that our new definition is not empty.

Adrian Kent

This doesn't necessarily mean that quantum money is the only possible form of money that satisfies the definition. We can't emulate the flexibility of quantum money with a single classical token, nor can we emulate the unforgeability of quantum money with money based on a single classical data string. But it turns out that we can reproduce both properties by schemes that use multiple classical data strings together with, for instance, variants of the secure relativistic quantum bit commitment protocol described above (Kent 2019). This may be fortunate, since quantum money requires a secure quantum memory for at least as long as the interval between issuing and spending, and long-term quantum memory is technologically very challenging. Even when (if) it becomes available, it will likely be expensive and cumbersome. Although these partly classical emulation schemes are also somewhat technologically challenging, they can be implemented in principle without quantum memory, using (at most) short range quantum communication together with long range classical communication. With experimental colleagues, we have begun test-of-principle implementations (Kent et al. 2021). I hope that we will be able to report further progress within the next year.

Might new laws of nature alter our understanding of physics-based cryptography?

I've discussed here what the presently known laws of nature, in particular quantum theory and special or general relativity, tell us is possible for communicating and controlling information. Cryptographers naturally worry about every assumption, so it's reasonable to ask just how much trust we should place in security based on current physical theory. This raises fascinating and fundamental questions in theoretical physics: Could there be as yet undiscovered ways of accessing physical information? What are the constraints? How much of existing theory would need to be modified if it were true, and how likely is this?

Cryptographers can take some reassurance from the fact that the relevant principles are cornerstones of very well-established theories. Special and general relativity define a causal structure of space-time, from which it follows that no form of information can travel faster than light. If that principle is broken, it seems that we run into causal loop paradoxes, in which signals sent into the past affect the present and prevent the signals from being sent. While there are theoretically possible ways to avoid these paradoxes, these tend to create other problems.

Quantum security is ultimately based on one of the fundamental postulates of quantum theory, namely that quantum measurements (which are how we extract classical information from quantum states) follow specific rules that constrain the amount and type of classical information that can be extracted. Moreover, some of the most straightforward ways of altering those rules – for instance by allowing unrestricted cloning of quantum states – would not only mean going beyond quantum theory, but also give a way of signalling faster than light, overturning special and general relativity as well.

Revolutions in fundamental physics have happened, of course, but rarely. Quantum theory and special and general relativity have been repeatedly confirmed, over a very wide range of domains, for around a century. Before that, Newtonian theories lasted for a couple of centuries. We might thus perhaps (very crudely!) estimate that the chance of these theories being overturned is no more than 1% a year – and likely less than that, given the enormously greater resources devoted over the last hundred years to developing fundamental science and to testing these theories.

It's probably also reasonable to assume that any hint of new laws of fundamental physics would be very hard to keep secret, so that any threat to physics-based cryptography would very likely come with advance warning. All in all, it seems unlikely that that cryptographic schemes with security based on quantum theory and special relativity could be broken in the next couple of decades. Indeed, it looks pretty plausible that these schemes will never be breakable because the basic principles of quantum theory and relativistic causality may well be retained in future unified theories. Still, since we cannot claim certainty, there is scope for reasonable concern for the small but significant subset of applications where very long-term secrecy is needed and the acceptable risk is very small. Even for those applications, though, physics-based cryptography can strengthen the security when combined with computationally based cryptography. This "belt and braces" approach gives schemes that can only be broken if new algorithms are found to attack presently computationally hard problems and the fundamental principles of quantum theory or special relativity turn out to be incorrect in a way that is cryptologically exploitable.

On the other side of the argument, there are deep unresolved problems in fundamental physics, some of which are closely connected to the question of how nature extracts information from quantum systems. Most obviously, we don't yet have a fully unified theory of quantum theory and gravity. As far as we understand it, matter is described by quantum theory, and in some sense nature uses information about quantum matter systems to produce gravitational fields. It's an open question whether gravity is fundamentally quantum – certainly on large scales gravitational fields appear as classical entities that we can measure classically. It's possible that nature uses different rules from those we currently understand to generate the classical information carried by gravitational fields from the quantum information carried by matter. It's even possible that these rules allow a form of quantum cloning that is consistent with relativistic causality, since it turns out (Kent 2005*b*, 2021) that one can define a reasonably natural variation of standard quantum cloning that does not imply faster than light signalling. Excitingly, new ideas for "table-top" experimental tests of whether gravity is quantum have recently been proposed (e.g., Bose et al. 2017, Marletto & Vedral 2017, Howl et al. 2021). These and related experiments (e.g., Kent 2021) may confirm the mainstream hypothesis that gravity is quantised and that the unified theory involves no novel form of information processing, but alternative results with more radical implications cannot be ruled out. The technology required is – once again! – challenging, but seemingly not completely out of reach, and it seems plausible that we may get experimental answers in the next decade or two.

Although it's much more speculative and controversial, another line of thought deserves mention. A respectable minority of physicists (e.g., Penrose 1994) and philosophers of mind (e.g., Chalmers 1996) argue that consciousness is another way in which nature extracts information (namely our thoughts and perceptions) from matter (our brains and nervous systems) by mechanisms unexplained by current physical laws. If so, there could be new physical laws that characterise the relationship between consciousness and the material world and that could also have new implications for the physics of information. It's probably fair to say that there are presently no very compelling theoretical proposals in this direction, and only relatively hazy ideas – for example, trying to create potentially conscious quantum simulations of agents on quantum computers – about how we might probe these questions experimentally. We might need completely new concepts to develop a physical theory of consciousness, if indeed there is one, and this might take centuries or even millennia.

In summary, from simple codes, seals and envelopes through to relativistic quantum cryptography, the story of how we closet and control information is intertwined with the development of theoretical physics and technology, and none of these stories is close to complete. A new understanding of how nature defines, controls and processes information would be scientifically extraordinarily exciting, quite apart from potential applications – but the possibility that new fundamental discoveries could enhance the power of quantum computing and quantum-assisted communication is also arguably one of the strongest reasons for practically minded funders to continue to support theoretical and experimental work in fundamental physics. Even if the probability of these spin-offs is relatively small, their potential impact is arguably large enough to justify significant investment. And perhaps in the long run this probability is not so small: we have learned, in more than one sense, that nature keeps secrets well.

Adrian Kent

References

Bennett, C. H. & Brassard, G. (1985), An update on quantum cryptography, *in* G. R. Blakley & D. Chaum, eds, 'Advances in Cryptology CRYPTO 1984, Lecture Notes in Computer Science', Vol. 196, Springer-Verlag, Berlin, Germany, pp. 475–480.

Blum, M. (1983), 'Coin flipping by telephone: a protocol for solving impossible problems', *ACM SIGACT News* **15**, 23–27.

Bose, S., Mazumder, A., Morley, G. W., Ulbricht, H., Toroš, M., Paternostro, M., Geraci, A. A., Barker, P. F., Kim, M. S. & Milburn, G. (2017), 'Spin entanglement witness for quantum gravity', *Physical Review Letters* **119**, 240401.

Brassard, G., Chaum, D. & Crepeau, C. (1988), 'Minimum disclosure proofs of knowledge', *Journal of Computer and System Sciences* **37**, 156–189.

Chalmers, D. J. (1996), *The Conscious Mind: In Search of a Fundamental Theory*, Oxford Paperbacks, Oxford, U.K.

Dieks, D. G. B. J. (1982), 'Communication by EPR devices', *Physics Letters A* **92**, 271–272.

Hayden, P. & May, A. (2016), 'Summoning information in spacetime, or where and when can a qubit be?', *Journal of Physics A: Mathematical and Theoretical* **49**, 175304.

Howl, R., Vedral, V., Naik, D., Christodoulou, M., Rovelli, C. & Iyer, A. (2021), 'Non-Gaussianity as a signature of a quantum theory of gravity', *PRX Quantum* **2**, 010325.

Kent, A. (1999), 'Coin tossing is strictly weaker than bit commitment', *Physical Review Letters* **83**, 5382.

Kent, A. (2005*a*), 'Nonlinearity without superluminality', *Physical Review A* **72**, 012108.

Kent, A. (2005*b*), 'Secure classical bit commitment using fixed capacity communication channels', *Journal of Cryptography* **18**, 313–335.

Kent, A. (2011), 'Unconditionally secure bit commitment with flying qudits', *New Journal of Physics* **13**, 113015.

Kent, A. (2013), 'A no-summoning theorem in relativistic quantum theory', *Quantum information processing* **12**, 1023–1032.

Kent, A. (2018), 'Unconstrained summoning for relativistic quantum information processing', *Physical Review A* **98**, 062332.

Kent, A. (2019), 'S-money: virtual tokens for a relativistic economy', *Proceedings of the Royal Society A: Mathematical, Physical and Engineering Sciences* **475**, 20190170.

Kent, A. (2021), 'Testing quantum gravity near measurement events', *Physical Review D* **103**, 064038.

Kent, A., Lowdnes, D., Pitalúa-Garcia, D. & Rarity, J. (2021), 'Practical quantum tokens without quantum memories and experimental tests', npj *Quantum Information* **8**, 1–14.

Landauer, R. (1991), 'Information is physical', *Physics Today* **44**, 23–29.

Lo, H. K. & Chau, H. F. (1997), 'Is quantum bit commitment really possible?', *Physical Review Letters* **78**, 3410.

Marletto, C. & Vedral, V. (2017), 'Gravitationally induced entanglement between two massive particles is sufficient evidence of quantum effects in gravity', *Physical Review Letters* **119**, 240402.

Mayers, D. (1997), 'Unconditionally secure quantum bit commitment is impossible', *Physical Review Letters* **78**, 3414.

Molina, A., Vidick, T. & Watrous, J. (2012), Optimal counterfeiting attacks and generalizations for Wiesner's quantum money, *in* 'Conference on Quantum Computation, Communication, and Cryptography', Springer, Berlin, Germany, pp. 45–64.

Penrose, R. (1994), *Shadows of the Mind*, Oxford University Press, Oxford, U.K.

Shannon, C. E. (1948), 'A mathematical theory of communication', *Bell System Technical Journal* **27**, 379–423.

Shannon, C. E. & Weaver, W. (1949), *The mathematical theory of information*, University of Illinois Press, Illinois, U.S.A., pp. 128–164.

Adrian Kent

Turing, A. M. (1936), 'On computable numbers, with an application to the Entscheidungsproblem', *Journal of Mathematics* **58**, 345–363.

von Neumann, J. (1945), 'First draft of a report on the EDVAC', https://archive.org/details/firstdraftofrepo00vonn/page/n3/mode/2up.

Wiesner, S. (1983), 'Conjugate coding', *ACM SIGACT News* **15**, 77–88.

Wootters, W. K. & Zurek, W. H. (1982), 'A single quantum cannot be cloned', *Nature* **199**, 802–803.

4 Antarctica: isolated continent

JANE FRANCIS
British Antarctic Survey

Abstract: Continents as we know them today emerged as a consequence of the mechanism of plate tectonics, which led to the fragmentation of a super-continent. One such fragment, the Antarctica, now is in the ocean at the South Pole, covered in thick ice-sheets that contrast with its long-past history where it was adorned by forests and inhabited by animals including dinosaurs. It was the natural processes that buried carbon dioxide that led to the glaciation of Antarctica. The burning of fossil fuels is now having an opposite effect, causing the depletion of the ice at a remarkable rate. For humans, Antarctica can be thought of as an isolated continent because no one actually makes a home there. But the continent is not entirely isolated – there is life, including a few thousand scientists and their support staff. And the oceans around are teeming with life with a few species of birds breeding on the continent.

INTRODUCTION

Antarctica is the coldest, driest, highest continent on Earth. It is also the most isolated, sitting over the South Pole and surrounded by the Southern Ocean, which includes the most continuous ocean current on the planet. Its current isolation over the South Pole strongly influences its landscape today - one of thick ice sheets, enormous ice streams, plus a rim of ice shelves and winter sea ice that link it to the surrounding cold polar ocean.

Antarctica's current isolation, along with its remoteness and forbidding polar climate, is relatively recent (geologically-speaking) but is an aspect of our planet that we have known throughout the period of human evolution. Many of the animals and plants now associated with the continent are endemic, found only in and around Antarctica, and are well adapted to the frigid climate. Humans can only survive in this hostile remote environment with the aid of modern technology which decreases the isolation.

Geological influence on connection, not isolation

The Antarctic continent, however, has not always been isolated. On a geological timescale it was a critical piece in the puzzle of tectonic plates that formed the great southern landmass of Gondwana. About 100 million years ago during the Cretaceous period (the time of the dinosaurs), movement of tectonic plates positioned Antarctica over the South Pole, where it has remained up to today. During the Cretaceous Antarctica was still joined to Australia and to South America, forming a continuous landmass that allowed land animals and plants to migrate from one continent to another (Figure 4.1).

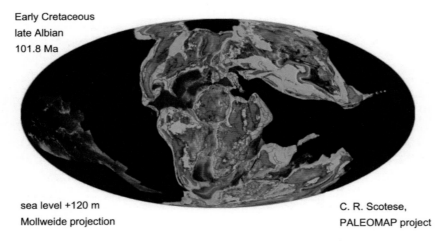

Early Cretaceous
late Albian
101.8 Ma

sea level +120 m
Mollweide projection

C. R. Scotese,
PALEOMAP project

Figure 4.1 Palaeogeographic reconstruction of the Earth during the Cretaceous, about 100 million years ago.

Global climates were much warmer too, warmed by natural greenhouse gases from volcanoes. Evidence for polar warmth comes from many plant fossils in rocks from Antarctica, particularly of Cretaceous age. Despite the current 99% ice cover of the Antarctic continent, rocks are exposed on mountain tops above the ice sheets and on ice-free islands around continental margins. Many of these rock sequences contain fossil leaves, fossil flowers, pollen and spores, and petrified tree stumps, which provide an insight into life on the Antarctic land mass in warmer times. The plant fos-

Jane Francis

sils have been identified as ancient ancestors of modern Southern Hemi-sphere vegetation, similar to that growing today in cold temperate regions like Patagonia, Tasmania and New Zealand. Plant fossils include those of monkey puzzle conifers (*Araucaria araucana*), podocarp conifers, Ginkgo trees, Proteacea shrubs, and most commonly, southern beech (*Nothofagus*).

Working with artists, my colleagues and I have produced pictorial re-constructions of Antarctic forests. Figure 4.2 is an oil painting by artist Robert Nicholls of an Antarctic forest that grew about 70°South approxi-mately 100 million years ago – the fossil forest is preserved today on ice-covered Alexander Island on the Antarctic Peninsula at approximately the same latitude. The forest is reconstructed as accurately as possible using fossil evidence of leaves, roots and plant remains, plus measurements of the spacing of fossil tree stumps on rock layers. Ginkgo trees, podocarp and monkey puzzle conifers were the dominant trees. In the undergrowth were tree-ferns, cycad-like plants, and small liverworts and mosses, along with some undergrowth plants that became extinct. Flowering plants were rare in Antarctica at that time. Estimation of the global CO_2 level at that time, using a range of geological proxies, suggest that the atmospheric CO_2 was probably over 1000ppm.

Figure 4.2 Oil painting reconstruction of 100 million year old forest that grew at 70°S, Antarctic Peninsula. The forest is reconstructed using data from PhD project of Jodie Howe and other geologists. Artwork by Bob Nicholls, Paleocreations. The painting is on display at the British Antarctic Survey.

From about 80 million years ago, flowering plants became an important component of the lowland forests of Antarctica, for example, southern beech (*Nothofagus*), Proteaceae shrubs (similar to modern *Banksia*), Winteraceae (*Drimys*), Gunnera and other waterside plants (Bowman et al. 2014). Monkey puzzle conifers and heathland plants grew up in volcanic highlands (a southern extension of the Andes), now represented by the Antarctic Peninsula. Fossil teeth and bones indicate that the forests were inhabited by a range of dinosaurs, including sauropods, theropods, hadrosaurs, ornithopods, plus birds (Figure 4.3). Even after the mass extinction event 66 million years ago, which saw the end of dinosaurs in Antarctica, the continent was still covered with lush forests in warm climates, inhabited by mammals, ungulates, carnivorous birds and the earliest penguins through until about 50 million years ago (Eocene). These plants and animals lived in the warm polar climates of Antarctica due to the connections across the southern continents, before the continent became isolated.

Figure 4.3 Reconstruction of 70 million year old Antarctic forest based on geological data. Monkey puzzle trees grew on the mountains and the lowlands were covered with southern beech, flowering Proteaceae, tree ferns, and waterside plants. Fossilised dinosaur bones indicate a range of dinosaurs lived in the forests. Artwork by James Mackay, Leeds (in Bowman et al. (2014)).

Jane Francis

Did continental isolation lead to antarctic glaciation?

Geological evidence of past climates from various geochemical, sedimentological and paleontological sources indicates that the south polar region began to cool about 45 million years ago. However, definitive evidence in the rock record for the earliest glaciers is unfortunately buried beneath the current thick ice cover on the continent.

Indisputable evidence for past glaciation on Antarctica was, however, discovered in the 1990s with the innovative development of drilling platforms that sat on thick sea ice at the margin of the continent and drilled rock cores from the strata on the sea floor below (Cape Roberts Drilling Project; AN-DRILL programme). The rock cores contained evidence of the transition from non-glacial to glacial conditions with the appearance of sediment types that provided clear evidence for the presence of an Antarctic ice sheet at least 40 million years ago. The cause of global cooling that allowed ice to form on Antarctica has since been shown by climate models to have been due to decrease in atmospheric CO_2 through natural processes that buried carbon in rock reservoirs on a geological time scale of millions of years ((Rae et al. 2021)).

For many years it was considered that the south polar region became glaciated as a result of cold climates caused by the tectonic movements that isolated Antarctica from the rest of the southern continents, although now we know that decreasing CO_2 was probably the main driver. Around 20 million years ago the last link to other southern continents was broken by tectonic activity, resulting in the formation of the deep channel that is now the Drake Passage between the tip of South America and the Antarctic Peninsula (Figure 4.4). This allowed ocean currents to flow unhindered around Antarctica for the first time, creating the stormy Circum-Antarctic Current and isolating the continent within deep cold waters of the Southern Ocean.

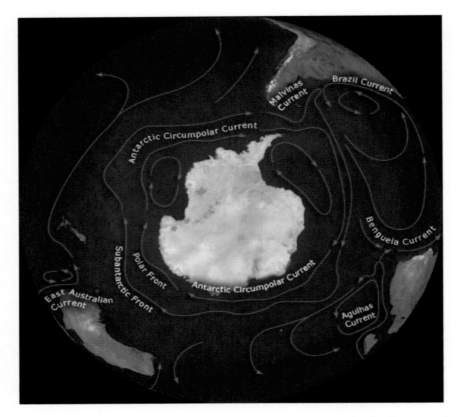

Figure 4.4 Antarctica became separated from other continents as tectonic plates moved apart. Antarctic Circumpolar Current flowed around Antarctica for the first time, isolating the continent in cold icy waters.

Jane Francis

Isolation and the growth of the ice sheets

The isolation of the continent within the Southern Ocean intensified the cooling, separating warmer tropical waters from cold ocean water in the south (called the Antarctic Convergence or Polar Front). Atmospheric CO_2 levels continued to fall (Rae et al. 2021). This ultimately led to the growth of massive ice sheets that extended way beyond the land area of Antarctica and covered the continent in ice sheets that today are over 4km thick in places. Global cooling eventually affected the whole planet, allowing ice sheets to form in the Arctic about 2-5 million years ago during the Pliocene epoch. By that time it is estimated that the global CO_2 level was about 400ppm, much like that of today (Figure 4.5).

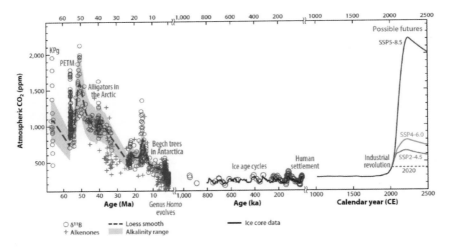

Figure 4.5 Atmospheric CO2 level through the past 66 million years from geological evidence (in Rae et al. (2021))

The massive ice sheets and intense cooling that isolated the continent had a drastic impact on Antarctica faunas and floras. Fossils of small bushes of southern beech, mosses, cushion plants and weevils are preserved within glacial deposits in the Transantarctic Mountains, only 500km from the South Pole today. They represent the last tundra vegetation that grew on Antarctica between 10-5 million years ago in an interglacial period when ice sheets retreated a little (Figure 4.6). The subsequent intense global cooling and extension of the ice sheets heralded the end of the ancient Antarctic vegetation. Today there are no trees or shrubs and the only plants that survive on the continent through the short summers and cold dark winters are mosses, liverworts and lichens, and two species of small flowering plants. Similarly, the dinosaurs and land mammals of the past disappeared, and only small arthropods such as mites and springtails live on land today.

Figure 4.6 Fossil leaves of southern beech (*Nothofagus*) from the Transantarctic Mountains (leaves about 1 cm in length). These small tundra bushes were the last vestiges of the past Antarctic forests which covered Antarctica before icesheets covered the continent.

The ocean around Antarctica is particularly rich in bird life, but only three – Emperor penguin, skuas and the Antarctic Petrel actually breed on the continent. Many live on the Antarctic coast and others, such as albatross, live amongst the rich feeding grounds of the sub-Antarctica (e.g. South Georgia). Antarctic ocean waters are particularly fertile near the Polar Front where cold Antarctic waters meet warm waters of the sub-Antarctic. Upwelling currents produce very high productivity of phytoplankton, which feeds the krill that themselves feed the food chain of fish, penguins, seals, and whales. Nearshore ocean ecosystems contain specialised corals, sponges, isopods, pycnogonids, starfish and more, many of which grow very slowly but to old age. Many of these marine animals are endemic (Clarke et al. 2005) and are well adapted to the stable but cold temperatures; for example, fish contain antifreeze in their veins instead of red blood cells. Will these special ecosystems adapt to warming waters or become extinct in future?

Jane Francis

Isolated but with global impact

Despite being the only uninhabited continent on Earth and being isolated in its remote and icy environment, Antarctica is not isolated from the impact of anthropogenic warming. The polar regions are the most sensitive to global change, as seen in the geological past – change happens first at the poles and most dramatically. Antarctica has recently experienced increased air temperatures of 3°C, especially around the Antarctic Peninsula, much greater than the average global rate. It is becoming clear that climate warming in the tropics causes strengthening winds that drive warm ocean waters under ice shelves fringing Antarctica, melting them from below. Many international collaborative projects in Antarctica are currently focused on exploration and observation of ice shelves and glaciers, such as the joint US NSF/UK NERC Thwaites Glacier Collaboration.

Ice shelves around the edge of the continent act as buttresses, keeping glaciers and ice streams on land, hence the concern is that as ice shelves melt and the buttress effect diminishes, glacier ice will flow faster into the ocean, ultimately raising global sea levels. The Thwaites Glacier, in the Amundsen Sea on the western coast of Antarctica, is in a region that is warming faster than other areas and satellite observations indicate the greatest thinning of the ice here. The Thwaites Glacier project is a multi-observation project to understand how warm water is melting the glacier from below, which threatens the stability of the West Antarctic Ice Sheet and could lead to global sea level rise of up to 3m. NASA have demonstrated the loss of Antarctica ice from 2002 to 2020 (NASA and JPL-Caltech 2021). Satellite data show that Antarctica lost 149 billion metric tons of ice each year from 2002 to 2020, adding to global sea level rise. Most ice is lost from western Antarctica from the West Antarctic Ice Sheet (Figure 4.7).

Global sea level is now rising at approximately 3-4mm per year, of which about 1-2mm is from melting ice sheets. Coastal regions of the world, with their large human populations, major cities and supply chains are all threatened by melting ice sheets on Antarctica. What happens in Antarctica now affects us all across the planet.

Human isolation on this isolated continent

What is it like to work in Antarctica so remote and isolated from the rest of the world? The long cold dark winters make it a difficult and dangerous place to be, without support from passing ships and accessible by aircraft only for dire emergencies.

There are no permanent inhabitants on the continent but each summer the research stations across the continent are home to about 4000 scientists and their support staff. Fifty four nations work in Antarctica, based in about 70 research stations, some as small single bases, others like large towns. Antarctic tourism is also growing rapidly, either on cruise ships on the Antarctic ocean or on land in a few expedition camps - in the 2019/20 season over 70,000 tourists visited the region.

So being in Antarctica during the summer is not so isolated as it was. Most stations have telephones and satellite communication, and internet access may be slow but it is possible to keep in touch as normal from an Antarctic station. Scientists who go into the depths of Antarctica for their field work are isolated in their tent camps and rely on good communications with the home station by radio or satellite phone. Despite living in relative isolation in a field camp, ironically, most people are always accompanied by at least one other person for safety, and that is usually for 24 hours per day. It is hard (and not safe) to be alone in Antarctica.

The winter in Antarctica is cold and dark and truly isolates the continent. Transport that can handle the extensive sea ice and cold icy conditions is normally called in only for life-saving emergencies. About 1000 people inhabit stations in winter, literally keeping the lights on and ensuring that some long-term data sets continue to be gathered.

This extreme winter isolation makes Antarctic stations ideal places for space experiments. The European Space Agency ESA has recognised that living in Antarctica is at times similar to living on another planet and have set up experiments to study the effect of isolation on small teams for long periods. ESA sponsored a medical research doctor in the French-Italian station called Concordia, which is one of the most remote and relatively high-altitude stations (3200m above sea level) near the South Pole, to study the long-term effects of isolation to inform them of human response to long flights aboard the International Space Station and voyages to Mars.

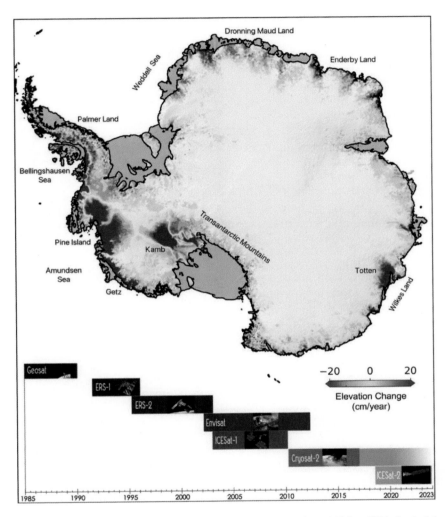

Figure 4.7 Changes in elevation of the Antarctic ice sheet from 1985 to 2021. Ice height diminishes (red) as the ice sheet melts by contact with ocean water; it rises (blue) where accumulation exceeds melting. Ice shelves are shown in grey. The satellite missions that supplied data are listed at the bottom. Thwaites Glacier is the larger red area next to Pine Island Glacier. Credit: NASA/JPL-Caltech (Rasmussen 2022).

ESA also set up experiments in UK's Halley station to simulate space flights. They were particularly interested in how humans can adapt to long space flights and whether they could remember skills taught many months ago. Halley personnel were taught how to dock a Soyuz spacecraft at the International Space Station (on a mock-up of the equipment) and tested regularly to see if the learnt skills diminished over time (British Antarctic Survey 2015).

Future isolation?

While the tectonic position of Antarctic is not likely to change for millions of years [modelled by Scotese (2015)], and the extreme seasonal changes in light from continuous summer sunlight to the long dark winter night near the South Pole will not change, our warming climate may affect the isolation of Antarctica.

As ice sheets melt, more land will be exposed to make available land-scapes for more verdant vegetation. Currently the Committee for Environmental Protection, part of the Antarctic Treaty, works hard to protect the Antarctic environment from alien species brought in by human visitors but warming climates may allow the survival and propagation of new species brought in by organisms such as birds and driftwood (Clarke et al. 2005). In the oceans the Polar Front and the dramatic change to cold ocean temperatures currently acts as a barrier to warmth-loving animals but with changing ocean circulation, influenced by temperature, we may see the endemic Antarctic icehouse faunas replaced by competitors from lower latitudes.

Our rapidly changing climate, caused by anthropogenic warming, may bring such dramatic changes to this southern icy polar landscape that Antarctic may no longer be isolated in future.

Jane Francis

References

Bowman, V. C., Francis, J. E., Askin, R. A., Riding, J. B. & Swindles, G. T. (2014), 'Latest cretaceous–earliest paleogene vegetation and climate change at the high southern latitudes: palynological evidence from seymour island, antarctic peninsula', *Palaeogeography, Palaeoclimatology, Palaeoecology* **408**, 26–47.

British Antarctic Survey (2015), 'Halley research station hosts research to understand human adaptation to space flight', *phys.org* .

Clarke, A., Barnes, D. K. & Hodgson, D. A. (2005), 'How isolated is Antarctica?', *Trends in Ecology & Evolution* **20**(1), 1–3.

NASA and JPL-Caltech (2021), 'Video: Antarctic ice mass loss 2002-2020', https://sealevel.nasa.gov/resources/120/video-antarctic-ice-mass-loss-2002-2020/.

Rae, J. W., Zhang, Y. G., Liu, X., Foster, G. L., Stoll, H. M. & Whiteford, R. D. (2021), 'Atmospheric co2 over the past 66 million years from marine archives', *Annual Review of Earth and Planetary Sciences* **49**(1), 609–641.

Rasmussen, C. (2022), 'Nasa studies find previously unknown loss of antarctic ice', https://www.nasa.gov/feature/jpl/nasa-studies-find-previously-unknown-loss-of-antarctic-ice.

Scotese, C. (2015), '240 million years ago to 250 million years in the future', https://www.youtube.com/watch?v=uLahVJNnoZ4.

5 Isolation of particles using optical tweezers

PHILIP H. JONES
University College London

Abstract: In 2018 Arthur Ashkin was awarded a half share of that year's Nobel Prize in Physics "for the optical tweezers and their application to biological systems". The work for which he was recognised had its origins more than thirty years before, and in the years since their invention, the uses of optical tweezers have grown far beyond biological systems, with numerous diverse applications across the chemical and physical sciences also. In this lecture we will look at the history of our understanding of the force that light exerts on matter, which has its origins in the observations of Johannes Kepler concerning the tails of comets. We will see how the concept of radiation pressure evolved from the work of James Clerk Maxwell, and trace its development to the experiments in which Arthur Ashkin first demonstrated the optical tweezers. Finally, we will examine just a few of the many uses of optical tweezers where their "light touch" and ability to trap a single microscopic particle and isolate it from its surroundings have proved invaluable.

Introduction: optical tweezers

An optical tweezer is an instrument that can isolate, hold and manipulate microscopic and sub-microscopic objects using a single, strongly focused laser beam (Jones et al. 2015). Their "light touch", and the remote, non-contact, damage-free nature of the trapping has led to their application in diverse fields across the physical, chemical and biological sciences. Figure 5.1 below demonstrates the versatility of the technique, both in terms of the variety of materials that can be trapped and the sizes of these objects, ranging from single atoms to cells of the order of a few micrometres.

In this essay, we will look at the history of research into the force that light can exert on matter, and the ideas that led to the invention of optical tweezers in 1986. We will then look at just a few of the ways in which optical tweezers have been used to isolate microscopic particles and thus control their position and apply carefully calibrated forces in a controlled way.

History of radiation pressure

The ability of light to exert a force on matter was recognised as early as 1619 by Johannes Kepler in his work *De Cometis Libelli Tres*, in which he ascribed the origin of comet tails to the action of the rays of the sun. However, it was not until the introduction of James Clerk Maxwell's theory of electromagnetism (Maxwell 1873) and its subsequent development by John Henry Poynting, a former student of Maxwell, in a series of publications starting with an article for the layman in the Unitarian periodical *The Inquirer* (Poynting 1903), that the phenomenon of radiation pressure was fully understood. A comprehensive account of Poynting's contribution to this subject can be found in (Loudon & Baxter 2012).

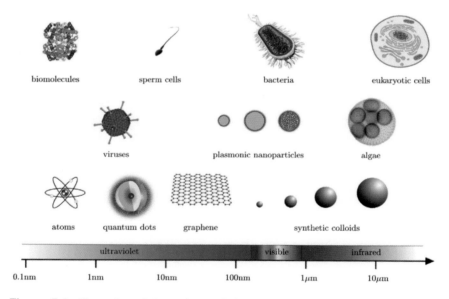

Figure 5.1 Illustration of the variety and sizes of objects that have been trapped in optical-tweezer experiments. Reproduced from (Pesce et al. 2020) under the Creative Commons CC-BY license.

Early experiments to detect the mechanical effects of light on Earth were

Philip H. Jones

performed by Nichols & Hull (1901), and separately by Lebedew (1901) at the very beginning of the twentieth century. These experiments used thermal light sources (electric or arc lamps), and focused the light onto a macroscopic mirror attached to a torsion balance, the deviation of the balance from its equilibrium position demonstrating the effects of radiation pressure. The forces involved were, inevitably, extremely small, leading Poynting to remark in his 1905 Presidential address to the British Physical Society:

> "A very short experience in attempting to measure these forces is sufficient to make one realize their extreme minuteness – a minuteness which seems to put them beyond consideration in terrestrial affairs"

It was not until the invention of the laser that it became possible to concentrate enough optical power in a small area to significantly affect the motion of microscopic objects. Much of the pioneering work in this area was performed by Arthur Ashkin.

Arthur Askin and optical trapping

Arthur Ashkin's name is synonymous with optical tweezers. He was born in Brooklyn, New York, in September 1922. Like his older brother, Julius, he studied physics at Columbia University and although his studies were interrupted by World War II, he graduated in 1947, and then later obtained a Ph.D. in nuclear physics from Cornell University in 1952. After completing his Ph.D. he joined Bell Laboratories, the research department of the telecommunications company AT&T, initially at Murray Hill, New Jersey, before moving to Holmdel, NJ (Essiambre 2021).

The Bell Laboratories have provided an extraordinary research environment, counting nine Nobel Prizes (and one Academy Award or 'Oscar')[1] for work that had its origins there, including the invention of the transistor, the discovery of the cosmic microwave background and, of course, Ashkin's award for the invention of optical tweezers. Mervin J. Kelly, Executive Vice President of the Laboratories in the 1950s put this productivity down to the freedom given to scientists to pursue and focus on their research work, even to the exclusion of fundamental development into basic technologies. As he wrote (Kelly 1950):

---

[1] A technical award in 1936 to E. C. Wente and the Bell Telephone Laboratories "for their multicellular high-frequency horn and receiver"

"The non-scientific duties of management should be minimised for
all levels of the research supervision ... There should be the very
minimum of diversion of the attention of the research leadership and
the individual researchers from their scientific programs."

The Laboratories also took a famously long-term view of research, sum-
marised by one researcher as "what you're doing might not be important for
ten years or twenty years, but that's fine, we'll be there then."

This dedication to research, even to that which may not necessarily seem
to be aligned with the needs of a telecommunications company, proved
highly fruitful for Ashkin. His early work included significant contributions
to nonlinear optics, such as the first observation of continuous wave second
harmonic generation (Ashkin et al. 1963), continuous wave parametric amplification (Boyd, 1966), and optical damage due to photorefractive index
modulation in LiNBO$_3$ (Ashkin et al. 1966). Ashkin, however, had a long-
standing interest in the mechanical effects of light, and reasoned that using
the newly-invented laser and microscopic particles, the forces that were dismissed by Poynting due to their "minuteness" could become large enough
that their effects could be seen. He made the following simple argument:

"Suppose we have a laser and we focus our one watt to a small spot
size of about a wavelength ≈1 μm, and let it hit a particle of diameter
also of 1 μm. Treating the particle as a 100% reflecting mirror of
density ≈ $1\,\mathrm{g\,cm^{-3}}$, we get an acceleration of the small particle =
$A = F/m = 10^{-3}\mathrm{dynes}/10^{-12}\,\mathrm{g} = 10^9\,\mathrm{cm\,s^{-2}}$. Thus, $A \approx 10^6\,g$,
where $g \approx 10^3\,\mathrm{cm\,s^{-2}}$, the acceleration of gravity. This is quite large
and should give readily observable effects. . ."

In 1970 Ashkin published the first in what would turn out to be a series
of papers leading to the demonstration of what we now know as "optical
tweezers" (Ashkin 1970). Here he demonstrated that microscopic spheres
suspended in water in a glass cell were drawn towards the axis of a laser
beam propagating horizontally through the cell, and then propelled at speeds
of several micrometres per second to be gathered at the wall of the cell.
Significantly, the forces acting on the particles could be explained through
the refraction and reflection of light alone, that is they were the effect of
radiation pressure rather than thermal gradients (radiometric forces), which
typically dominated over optical forces in earlier experiments.

Let us now consider the origin of the optical forces that Ashkin observed,
with reference to Figure 5.2, reproduced from (Ashkin 1970).

The laser beam, with a Gaussian intensity profile, is shown propagating

86

Philip H. Jones

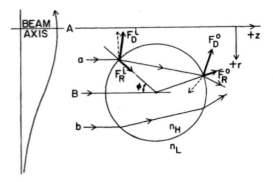

Figure 5.2 Ashkin's diagram illustrating the trajectory of a ray across a microscopic spherical particle that give rise to the gradient and scattering forces. The terminology is explained in the text. Reprinted figure with permission from A. Ashkin (1970). Copyright 2022 by the American Physical Society.

from the left of the figure in the direction $+z$, and incident on a sphere with refractive index n_H that is higher than the index n_L of the medium (water) in which it is suspended. The pair of rays labelled 'a' and 'b' are incident on the sphere symmetrically about the sphere axis (labelled B in the figure). The ray 'a' lies closer to the axis of the beam, labelled A, and so is stronger (more intense). Due to the refractive index difference between the water and the particle, it acts as a lens, hence on reaching the sphere a small fraction of the ray undergoes (Fresnel) reflection at the surface, whereas the larger portion is transmitted across the surface to the interior of the sphere, but its direction is changed by refraction. This gives rise to two components of the radiation pressure force at this surface from the ingoing ray: a force from the reflection of the ray, F_R^i, and a force from the deflection (refraction) of the ray, F_D^i. The ray crosses the sphere, and again when it next encounters the surface there are force components from this outgoing ray F_R^o and F_D^o from (Fresnel) reflection and refraction (deflection).

As Ashkin noted, all the forces arising from reflection and deflection at both surfaces contribute to an overall force in the direction of propagation of the beam $+z$. In later works this would come to be referred to as the *scattering force*, which propels a particle in the direction of the beam as Ashkin observed. However, they also contribute to an overall force that is *perpendicular* to the direction of propagation. In the case of the reflection forces these approximately cancel out, but in the case of the deflection (refraction)

87

forces these are directed along the radius of the beam towards its axis.

Now we must also consider the forces that arise from the reflection and deflection of the weaker ray 'b' (forces not marked on the diagram) that is symmetrically located about the axis of the sphere. By a similar argument we can see that reflection and deflection of this ray again contributes a scattering force along the direction of propagation $+z$, and also a resultant force that is perpendicular to this. In this case, however, due to the direction of refraction the radially directed force is *away* from the axis of the beam, opposing the force from ray 'a'. Since the ray 'b' is weaker, this outward-directed force is smaller than the inward-directed force from the more intense ray, 'a', and so the resultant effect is a force that is directed perpendicular to the direction of propagation of the beam, and acts to pull the particle towards the beam axis. As this force has its origin in the differing intensities of the rays that are incident on the particle because of the spatial gradient of the intensity of the beam, it later became known as the *gradient force*.

In this work, therefore, Ashkin explained the motion of the particles he observed: they were drawn to the axis of the beam where the intensity was greatest by the gradient force, then propelled along it by the scattering force. Furthermore he clearly demonstrated that these were optical forces, rather than thermal (or radiometric) forces caused by temperature gradients since the particles used were transparent and suspended in water. In the same short paper, Ashkin demonstrated the stable trapping of microscopic particles between two counter-propagating beams which balances the pushing effect of the scattering force from each beam; he also reported observations of optical forces on water droplets in air and gas bubbles in glycerol solution (where the direction of the gradient force is reversed to point away from the high intensity part of the beam, since the bubble has a lower refractive index than its surroundings). He further extended his ideas on the utility of optical forces to objects in vacuum, considering the acceleration and rotation of particles, and also the effect on an atomic vapour. Despite this, the paper very nearly never saw the light of day. In the Bell Laboratories' internal review process prior to submission, the report came back that the manuscript had "no new physics, and, even though there was nothing wrong with it, it was not worthy of Physical Review Letters."

Following the publication of this paper Ashkin explored numerous applications of optical forces (for Ashkin's own account of his work in this time see (Ashkin 1980, 1997, 2000). Notably he made significant contributions towards the laser cooling and trapping of neutral atoms (Ashkin &

Gordon 1979) which lead to the demonstration of optical trapping of laser cooled atoms (Chu et al. 1986), and the first magneto optical trap (Raab et al. 1987), both of which were realised at Bell Laboratories, and for which subsequently Steven Chu won a share of the 1997 Nobel Prize in Physics. He also continued to experiment with optical forces on microscopic objects, demonstrating, e.g., 'optical levitation' (Ashkin & Dziedzic 1971), where the upwards scattering force of a vertically directed laser beam was balanced by the downwards force of the particles weight. It was in 1986, at the age of 63, when many might be considering retirement, that Ashkin published the seminal paper that demonstrated for the first time a fully three-dimensional optical trap using a single beam, where confinement was produced by optical gradient forces alone, i.e. it did not rely on balancing scattering forces from counter-propagating beams, or balancing scattering forces with the particle weight. It is this trapping configuration that is now what is commonly known as the optical tweezers.

Ashkin's great insight that led to the development of the single beam optical trap (optical tweezers) was that a strong gradient of intensity along the direction of propagation of the beam could be created by focusing, and that if the focusing was strong enough the backwards-directed gradient force so generated could exceed the forwards-directed scattering force, and a microscopic particle could be trapped and isolated at the focus of the beam (Ashkin et al. 1986). In his ray optics model this axial force has its origin in the highly convergent rays coming from the edge of the aperture, the refraction of which results in a net force, the transverse components of which cancel out, and the remaining axial component is directed towards the geometrical focus. In this paper, Ashkin and his colleagues succeeded in trapping particles with sizes ranging over three orders of magnitude, from 10 μm down to 25 nm. The paper remains the most highly cited article in the journal Optics Letters.

Although the use of optical forces for trapping of atoms had been one of Ashkin's primary interests, he was quick to realise the opportunities the new device presented for trapping biological material. Very shortly after this first demonstration with synthetic colloids, he was was able to trap viruses and bacteria using visible laser light (Ashkin et al. 1987a). The method was further refined by the use of infra-red laser light which minimises optical damage to the trapped cell Ashkin et al. (1987b). Ashkin was thus able to trap and manipulate both whole cells (bacteria, protozoa, yeast, red blood cells), and sub-cellular structures (chloroplasts within green algae, organelles within protozoa) without damage Ashkin & Dziedzic (1989). In

2018, more than 30 years after the first optical tweezers experiment, Arthur Ashkin was awarded a half share of the Nobel Prize in Physics,[2] with the citation

"for the optical tweezers and their application to biological systems."

At the age of 96, Ashkin became the oldest Nobel Laureate at the time of the award.[3]

Optical tweezers: experimental methods

Figure 5.3 illustrates a simple optical tweezers experiment. Figure 5.3a shows a photograph of the instrument: essentially a home-built inverted microscope with the addition of a laser for trapping. A time-lapse video of the construction of this optical tweezer can be seen in the Supplementary Information of (Pesce et al. 2015). Figure 5.3b illustrates the components on plan view. Figures 5.3c-e demonstrate simple optical trapping and manipulation: a 1 µm diameter polystyrene sphere (circled), initially resting at the bottom of the sample, is trapped in the optical tweezers and isolated from the other particles. As the microscope stage is moved vertically the trapped particle remains in focus in the trapping and image plane, while the others move out of focus. The microscope stage can now be freely translated in the horizontal plane while the isolated particle remains trapped. The particle will remain in the trap provided it is not disturbed by forces (e.g., viscous drag forces) greater than the maximum that the optical tweezers can provide, typically of the order of a few tens of piconewtons.

Although the trapped particle is isolated from other particles in the sample, it is not completely isolated from the environment. As a colloidal particle suspended in a fluid it continually undergoes Brownian motion, even while held in the optical tweezers. As a result it is never completely still, but undergoes small fluctuations in position about its equilibrium position. These fluctuations provide a route for experimentally characterising the strength of the forces exerted by the optical tweezers, thereby turning them into a useful tool as a force transducer on the piconewton scale. When the trapped particle is disturbed from its equilibrium position it experiences a restoring force that pulls it back towards equilibrium, the magnitude of

---

[2]The other half of the prize was shared by Gérard Mourou and Donna Strickland "for their method of generating high-intensity, ultra-short optical pulses", namely chirped pulse amplification.

[3]This record was surpassed the following year, when John B. Goodenough was awarded the Nobel Prize in Chemistry at the age of 97.

Philip H. Jones

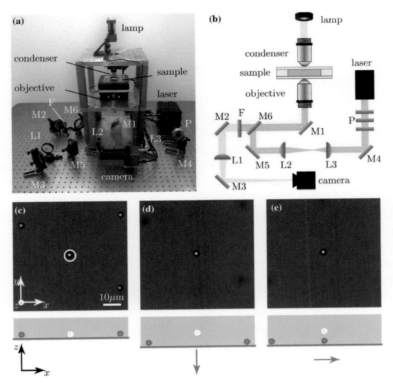

Figure 5.3 A simple optical tweezers experiment. (a) Photograph of the optical tweezers, with components labelled. (b) Schematic plan of the apparatus (M1-M6, mirrors; P, polarization optics; L1-L3, lenses; F, short-pass filter). (c) The circled particle is trapped in the optical tweezers. (d) The stage is moved vertically, and the trapped particle remains in focus; (e) The stage is moved horizontally, and the particle remains trapped. Figure reproduced from (Pesce et al. 2015). Copyright 2022 Optica Publishing Group.

the force being (to a good approximation) proportional to the displacement of the particle. This behaviour is familiar to all undergraduate students of physics as Hooke's Law, describing the force that acts when a spring is extended. The characteristic 'stiffness' of the spring is described by the constant of proportionality between the displacement and the force. For a microscopic particle suspended in water, however, the Reynolds number is extremely low, meaning that inertial effects are negligible (Purcell 1977) and the spring is strongly overdamped.

For such a mass-spring system undergoing thermal fluctuations the motion has particular characteristics, first studied by (Uhlenbeck & Ornstein 1930). These characteristics provide the route to calibration: an analysis of the trapped particle trajectory by any one of several techniques (e.g. mean-squared displacement, position autocorrelation, power spectrum analysis) and fitting the result to the known form yields a fit parameter that depends on the (known) hydrodynamic properties of the particle, and (the to be determined) trap stiffness (Jones et al. 2015). Typically the trap stiffness will depend on the properties of the particle (size, shape, material, refractive index) and also the properties of the focused laser beam being used for trapping (wavelength, diameter, numerical aperture of focusing). The calibration process is highly significant for optical tweezers, since following calibration the force acting on the trapped particle can be determined by simply measuring the displacement of the particle in the trap and multiplying by the spring constant. It is this ability to exert precisely calibrated forces on the piconewton scale with femtonewton resolution that makes the optical tweezers such a powerful device.

Optical tweezers: theoretical methods

Our understanding of the trapping mechanism for optical tweezers can be based on the action of ray optics, as outlined in (Ashkin et al. 1986), and further expanded in (Ashkin 1992). The ray optics methods is a good approximate method for calculating the force exerted in the size regime where the particle is much larger than the wavelength of the trapping light. In the contrasting regime where the particle is much smaller than the wavelength an alternative approximation, where the particle is treated as a point dipole is appropriate.

In this dipole approximation, the particle is considered to be polarisable by the electric field of the trapping laser beam, which induces a dipole moment that is proportional to the electric field. This induced dipole interacts with the electric field that induced it, and hence the dipole interaction energy is proportional to the square of the electric field, and thus the intensity. A particle that has a higher polarisability than its surroundings thus has its potential energy lowered in regions of high intensity, and experiences a force that pulls it towards those regions that is proportional to the gradient of the potential, and so the gradient of intensity. In a focused Gaussian laser beam the potential is approximately harmonic, and so the trapping force varies linearly with displacement (i.e. Hooke's Law).

Both approximations, therefore, in regimes of contrasting size of the particle describe the same qualitative behaviour: particles that have a higher refractive index (are more polarisable) than their surroundings experience a gradient force that confines them to the region of highest intensity, and that this force varies linearly for small displacements from equilibrium, like a Hookean mass on a spring. The vast majority of optical tweezers experiments, however, use particles that are a few micrometres in size, and laser light in the infra-red part of the spectrum, that is, with a wavelength of about one micrometre. Such experiments, therefore, do not strictly lie in either of these extreme size regimes, and it is only more recently that methods based on electromagnetic scattering theory have been applied for particles in this size regime.

In brief, the electromagnetic scattering problem is to find the electromagnetic fields scattered by a particle when illuminated by an incoming plane wave, illustrated in Figure 5.4a. In general, for an incident plane wave we are required to find the electromagnetic field inside the scattering particle and the electromagnetic field scattered by the particle such that the boundary conditions for continuity of the field components across the particle's surface are satisfied. The solution to this problem for scattering of a linearly polarised plane wave by a uniform homogeneous sphere was first obtained by Gustav Mie (1908). The incoming, internal and scattered fields are expanded in terms of vector multipoles, and continuity at the surface used to find the amplitudes in the expansion of the scattered field. More generally, the incoming and scattered fields can be related by a linear operator, **T**, the transition matrix that relates the coefficients of the multipole expansions, and contains all information about the size and orientation of the particle (Borghese et al. 2007). The **T**-matrix method is also suitable for calculating the scattering by non-spherical particles, particularly when modelled as aggregates of spheres. Other light-scattering methods that have been successfully applied to optical trapping calculations include the discrete dipole approximation, or coupled dipole model (Purcell & Pennypacker 1973), and the finite-difference time domain method (Gauthier 2005), both of which have the advantage of being applicable to particles of arbitrary shape and composition, but can be computationally demanding.

The scattering process redistributes the electromagnetic field of the incoming beam, implying an exchange of momentum between the field and the particle and hence a force acting on the particle. To find this force, the flux of momentum entering and leaving a surface surrounding the particle must be found, for which the form of the scattered field using one of the methods above is required. The momentum flux density of the field is obtained by forming the quantity known as the Maxwell Stress Tensor from the incident and scattered fields, the integral of which over the surface equates to the force acting on the particle.

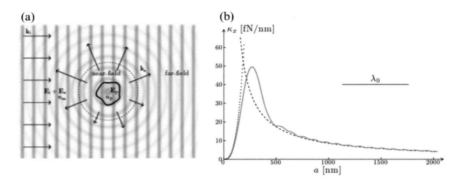

Figure 5.4 (a) Illustration of the general scattering problem: a particle scatters an incoming plane wave. The internal and scattered fields that satisfy the boundary conditions on the surface of the particle must be found; (b) comparison of approximate and exact methods for calculating optical trap stiffness across sizes. Dotted line – dipole approximation; dashed line – ray optics approximation; solid, red line – electromagnetic scattering theory. Figures reproduced from Jones et al. (2015), Cambridge University Press.

Figure 5.4b illustrates a comparison between the approximate methods and the full electromagnetic scattering approach by plotting the trap stiffness as a function of particle size for a fixed wavelength of light. The dotted line is the dipole approximation that is valid for small particles, and the dashed line the ray optics approximation (valid for large particles). The solid line is the full electromagnetic scattering calculation, which shows good agreement with the approximations in the extreme size ranges, but both approximate methods deviate from this in the intermediate regime where the size of the particle is comparable to the laser wavelength.

Applications of optical tweezers

As mentioned above, the impact of Ashkin's invention has been felt well beyond the Physics discipline boundaries. The ability to isolate microscopic and sub-microscopic materials has proved invaluable across diverse fields. The range of applications is so huge that we cannot consider all of them

Philip H. Jones

here, and the interested reader is referred to, e.g., (Jones et al. 2015, Polimeno et al. 2018, Gieseler et al. 2021). We will select just a few of these many topics, and start with applications in biology, as included in Arthur Ashkin's Nobel Prize citation.

Applications in biological sciences

Optical tweezers are particularly suited to studies of the physical properties of single biological molecules, due to the overlap in characteristic length, from nanometres to micrometres, and force (from femtonewtons to piconewtons) scales between biomolecules and optical tweezers. Mostly, biological molecules are not trapped directly, but attached to the surface of microbeads that act as a handle for the optical tweezers. The elastic properties of DNA are a classic example of the application of optical tweezers to single molecule biophysics. For small extensions, the mechanical properties of DNA can be described by the so-called worm-like chain (WLC) model (Marko & Siggia 1995), where elasticity is dominated by entropic contributions. Pioneering experiments by (Smith et al. 1996) used a two bead geometry with a micropipette and optical tweezers to stretch double stranded DNA (dsDNA) beyond the small extension regime. In these experiments the dsDNA length could be measured from the separation of the beads on the micropipette and in the optical trap, and the applied force from the displacement of the bead in the calibrated optical tweezers. For applied forces below 5 pN followed the WLC model, but for larger forces (over 10 pN) the force-extension curve deviated from these predictions, indicating the significance of the enthalpic contribution to elasticity, and requiring the modification of the WLC to include an elastic modulus (Wang et al. 1997).

In addition to linear stretching, the torsional properties of DNA can also be probed using optical tweezers. One way to achieve this is with a novel variation on the standard optical tweezers known as an 'optical torque wrench' (Porta & Wang 2004). This technique utilises the transfer of spin angular momentum between the polarised trapping laser beam and an optically anisotropic (birefringent particle). Application of torque while maintaining the DNA molecule under tension allowed measurement of the torsional modulus of DNA (Deufel et al. 2007), and also a transition from the usual (B-DNA) form to a high helicity, supercoiled (P-DNA) form. Intriguingly, for small applied torque an extension of the molecule was observed, explained by a twist-stretch coupling in the molecule (Gore et al. 2006).

In addition to single molecules optical forces can be applied to whole cells. (Agrawal, Smart, Norbe-Cardoso, Richards, Bhatnagar, Tufail, Shia, Jones & Pavesio 2016) used a dual optical tweezers method to study the elasticity of red blood cells (RBCs) at the whole cell level. In these experiments the RBCs were held in two optical tweezers and stretched by increasing the separation between the two traps, Figures 5.5a-c. Using blood samples from patients with Type 2 Diabetes Mellitus compared to control (healthy) patients, a correlation was found between the deformability of the cell subjected to this stretching procedure and Type 2 Diabetes, suggestive of a link between cell stiffness and diabetic retinopathy (Agrawal, Sherwood, Chhablani, Ricchariya, Kim, Jones, Balabani & Shima 2016). Similar RBC deformability measurements with optical tweezers have been applied to other conditions affecting cell stiffness such as infection by the malaria parasite (Suresh et al. 2005), and damage caused by radiation therapy (Inanc et al. 2021).

Alternatively, mechanical properties of localised areas of the cell, for instance a region of the cell membrane, can be measured using a microbead handle to pull a membrane tether, as in Figure 5.5d. From the applied force (deduced from the bead displacement) and the extent of the tether, the membrane stiffness can be determined. Belly et al. (2021) used this method to measure membrane stiffness of pluripotent stem cells, elucidating the signalling mechanism that regulated the decrease in membrane tension required for subsequent shape change at cell differentiation. The membrane tether pulling method has also been applied to synthetic vesicles by using optical tweezers to steer a pair of vesicles into contact before separating them, creating a tether that is formed by material from both vesicles, but is closed and does not permit transfer of material between the vesicles (Bolognesi et al. 2018), shown in Figure 5.5e. In further experiments mixing of the contents of initially separate vesicles was initiated by using optical tweezers to fuse the vesicles by heating gold nanoparticles on the vesicle surface. The resulting biomimetic microreactor system was used for controlled biochemistry by fusing three vesicles containing the components required for protein synthesis, leading to expression of green fluorescent protein (GFP) in the fused system, shown in Figure 5.5f.

Philip H. Jones

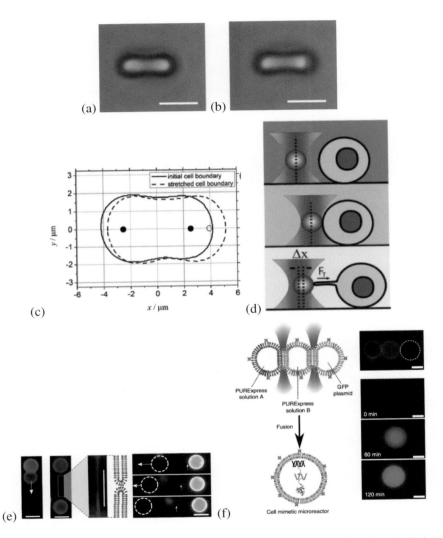

Figure 5.5 Applications of optical tweezers in biological sciences. (a) Red blood cells is held in dual optical tweezers (scale bar 5 μm); (b) Cell is stretched by increasing the separation between the traps. (c) Measured contours of the cell showing the deformation. (d) Principle of membrane tether pulling using optical tweezers; Reprinted from (Belly et al. 2021) under Creative Commons License CC-BY. (e) Pulling membrane tethers between synthetic vesicles. The fluorescence images show that there is no transfer of material between vesicles. (f) Optical tweezer induced vesicle fusion causing mixing of contents and expression of GFP in the fused vesicle; Reproduced from (Bolognesi et al. 2018) under Creative Commons Attribution 4.0 International License.

Applications to nanotechnology

In the first demonstration of optical tweezers, Ashkin et al. (1986) was able to trap dielectric particles that were tens of nanometres in diameter. Svoboda & Block (1994) were able to trap metallic nanoparticles of a similar size with a force that was some seven times greater than latex spheres, the difference being due to the seven times greater polarisability. Metal nanoparticles are resonant systems, and their optical properties are determined by plasmon resonances that are a function of particle size, shape and material Maier et al. (2007). Choosing a trapping laser wavelength that is on the long wavelength side of the plasmon resonance can enhance trapping times due to the increased real part of the polarizability, and hence gradient force. However the plasmon resonance is also associated with an increased extinction cross section, and hence scattering force which destabilises the trap. The ability to trap metallic nanoparticles can depend on the particle scattering properties determined by its size and shape (Pauzauskie et al. 2006, Yan et al. 2012)).

Carbon in the form of nanotubes and graphene may be amenable to optical trapping (Maragò, Gucciardi, Bonaccorso, Calogero, Scardaci, Rozhin, Ferrari, Jones, Saija, Borghese et al. 2008). For bundles of carbon nanotubes, the elongated shape not only increases trap stability compared to a nanoparticle with the same transverse dimensions, it also controls the alignment of the bundle in the optical tweezers, since it experiences a strong restoring torque to keep the nanotube bundle aligned with the trap axis (Maragò, Jones, Bonaccorso, Scardaci, Gucciardi, Rozhin & Ferrari 2008). The spatial distribution of position fluctuations, (Figure 5.6a) show that despite tight transverse confinement, the elongated nanostructure is weakly trapped in the axial direction, particularly when compared to a spherical particle in Figure 5.6b. It has been suggested that the small axial trap stiffness combined with the strong alignment and small transverse fluctuations, even at low trapping power, could make such optically trapped nanotubes a force sensor with femtonewton sensitivity. For graphene the optical trapping behaviour is further complicated by the strongly anisotropic refractive index, which has a large imaginary component when the light polarisation lies in the plane of the graphene. Graphene particles are only stably trapped, therefore, in an orientation with the plane perpendicular to the laser polarisation due to their optical anisotropy. This is in contrast with a flat microparticle with no optical anisotropy, which would be trapped with the plane parallel to the polarisation because of its shape (Maragò et al. 2010). The orientation

Philip H. Jones

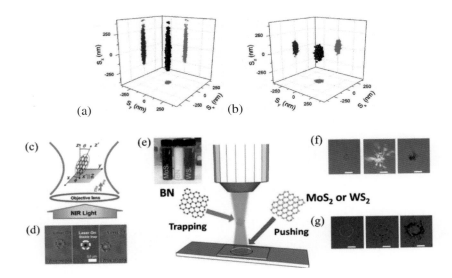

Figure 5.6 Optical trapping of nanomaterials. (a) 3D position tracking of trapped carbon nanotube bundle, showing the fluctuations along the laser propagation direction, and (b) of a spherical particle; Reprinted with permission from (Maragò et al. 2008). © 2022 American Chemical Society. (c) Trapping geometry for a graphene flake in optical tweezers; (d) images from experiment on trapping of graphene; Reprinted with permission from (Maragò et al. 2010) © 2022 American Chemical Society. (e) Metal dichalcogenides not trapped but propelled along the axis of the beam. (f) Set-up for trapping and optical force positioning 2D nanomaterials. (g) Optical force positioning on a substrate using a Gaussian beam. (h) Patterning deposition on a substrate using a ring-shaped beam. Reproduced from (Donato et al. 2018) with permission from the Royal Society of Chemistry.

of the graphene flake with respect to the trap propagation and polarisation directions is shown in Figure 5.6c, and Figure 5.6d shows a series of images of a graphene flake being trapped and released by optical tweezers.

Other two dimensional materials may have radically different optical trapping behaviours due to their optical properties. Hexagonal boron nitride (hBN), for example, is amenable to optical trapping using infra-red light, which unlike graphene flakes, is trapped with the plane parallel to the polarisation (Donato et al. 2018). The metal dichalcogenides MoS_2 and WS_2, however, are not trapped and are propelled along the axis of the beam. These differing behaviours are illustrated in Figure 5.6e. Interestingly this provides a method for patterning a substrate with these materials in a controlled manner, since the particles are confined within the high intensity parts of the beam profile, shown in Figures 5.6f and 5.6g for Gaussian and ring-shaped optical beams, respectively.

Applications to non-equilibrium physics

Optical tweezers also provide an ideal environment for the study of microscopic non-equilibrium systems. Such systems exhibit rich and complex dynamics due to the competition between an external driving and the internal response of the system that tries to oppose this to maintain equilibrium. One such system that shows fascinating dynamics is the driving of a colloidal microparticle across a spatially periodic potential created by an array of optical traps (Juniper et al. 2015). In this case driving with a time-periodic force leads to a synchronisation phenomenon termed dynamic mode locking. Since synchronisation phenomena occur widely in the natural world, from the coupled pendulum clocks studied by Huygens, to the flashing of fireflies, to asset price movements, it is a topic of broad interest. In the optical tweezers experiments the particles are driven through a periodic optical landscape by an external force which has both a constant and an oscillating component. Dynamic mode locking occurs when the frequency of modulation of the force exerted by the potential due to the motion of the particle being across it by the constant component of force becomes 'locked' to that of the oscillatory component of the external force. This is manifested by the appearance of 'Shapiro steps' in measurements of the particle average velocity as a function of the constant force amplitude, shown in Figure 5.7a. On the Shapiro steps the average velocity of the particle remains constant over a range of driving velocities as the mode-locked particle always hops back and forth the same number of potential wells in one cycle of oscillation of the force (Juniper et al. 2017).

Another phenomenon of a driven macroscopic system that can be studied on the micro-scale in optical tweezers is that of dynamical stabilisation. The paradigm of this is the Kapitza, or inverted pendulum, consisting of a weight (the pendulum bob) on a rigid rod, where the pivot of the rod is driven at high frequency in the vertical direction. This leads to the curious effect whereby the pendulum becomes stable in the inverted configuration. This phenomenon was first noted by Stephenson (1908), and explained by Kapitza (1951) by separating the timescales of the fast driving of the pivot and the slow oscillation of the pendulum. On the microscale the pendulum is realised by a ring-shaped optical trap, with fluid drag in the plane of the trap confining the particle analogous to gravity (Richards et al. 2018). Driving

Philip H. Jones

is provided by oscillation of the trapping potential parallel to the direction of fluid drag, illustrated in Figure 5.7b. Unlike the macroscopic system, in the microscopic case the system is strongly overdamped and exhibits richer dynamics, depending on the strength and frequency of the driving relative to the friction. Indeed, the fully inverted case of the (underdamped) macroscopic pendulum is unachievable, but stability can be found at any location up to the semi-inverted position depending on the details of the driving and the frictional environment, providing that the amplitude of driving is above a threshold, as shown in Figure 5.7c.

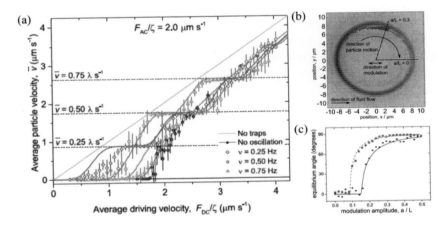

Figure 5.7 Microscopic dynamics of driven colloidal particles. (a) Dynamic mode locking as evidenced by 'Shapiro steps' in the average particle velocity as a function of the average driving velocity; reprinted from (Juniper et al. 2017) under Creative Commons License 3.0. (b) Particle trajectories evidencing dynamical stabilisation, superimposed on an image of the ring-shaped potential. Red data: with no driving the particle is pushed to the right by fluid flow. Blue data: with high frequency driving (above threshold) the particle stabilises at an angle above zero (but less than 90°). (c) Data demonstrating continuous control of the position of dynamical stabilisation for two different driving frequencies. Reproduced from (Richards et al. 2018) under Creative Commons Attribution 4.0 International License.

Alternative forms of trapping

Since the invention of optical tweezers many other forms of manipulation techniques for microparticles have been realised. Magnetic tweezers use magnets to create a strong field gradient to exert force on a superparamagnetic bead (Smith et al. 1992). These may be advantageous in some cases as the interaction of the magnetic field is only with the target probe particle, and typically does not affect the rest of the sample. Optoelectronic tweezers use light patterns to trap large numbers of particles in two dimensions (Wu & Ming 2011). The device consists of a transparent electrode and a photo-conductive electrode onto which patterns of light are projected and an electric bias applied which generates 'virtual electrodes' and dielectrophoretic traps in the illuminated areas. They have the advantage that they require low optical power and can use incoherent light sources such as LEDs to manipulate large numbers of particles, but they necessarily only act in two dimensions.

Optical binding is another two dimensional trapping method that occurs at an interface. Here light is focused at the surface of a prism at the critical angle for total internal reflection at a glass-water interface, shown in Figure 5.8a. A sample of microparticles dispersed in water is placed on the surface, and an evanescent optical field penetrates a short distance into the sample. Microparticles resting on the surface are immersed in the evanescent field and multiple scattering between the particles can lead to self-organisation of large numbers of particles in an ordered structure (Han et al. 2016). Figure 5.8b shows such a structure that forms in the interference pattern created by two pairs of counter-propagating evanescent fields. Not only electromagnetic fields can be used for particle isolation and manipulation, but also acoustic fields. This can be in an acoustic standing wave in a microfluidic device (Memoli et al. 2017), or with the acoustic analogue of optical tweezers using a single strongly focused beam (Baresch et al. 2016). Due to the sign of the acoustic radiation force, solid elastic particles are repelled from high intensity regions (in contrast to optical trapping), so a phased transducer array is used to create a vortex beam with zero amplitude on axis for trapping, shown in Figure 5.8c. Such acoustic tweezers can exert forces over a very different scale to optical tweezers, capable of applying micronewton forces to objects in the submillimetre size regime. Particles ranging in size from tens to hundreds of micrometres can be trapped, as shown in Figure 5.8d-f.

Philip H. Jones

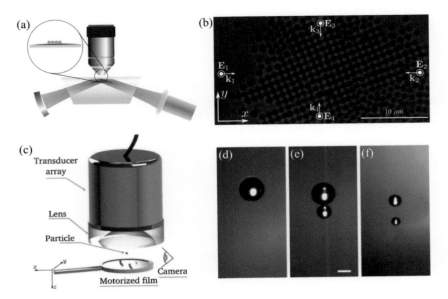

Figure 5.8 Alternative trapping schemes. (a) Experimental set-up for realising optical binding in an evanescent optical field. (b) Colloidal lattice of 1 μm diameter silica spheres formed by 2D optical binding. Reproduced with permission (Han et al. 2016). Copyright 2022 Optica Publishing Group. (c) Experimental apparatus for an acoustic tweezers; (d) trapped particle radius 190 μm; (e) trapped particles radii 177 μm and 108 μm. (f) Trapped particles radii 98 μm and 67 μm. Reprinted figures with permission from (Baresch et al. 2016) Copyright 2022 by the American Physical Society.

Outlook

While tremendous progress has been made with optical tweezers experiments, there remain many challenges. To cite just one example, in the field of biological applications considerable insight has been gained from *in vitro* experiments on single molecules and cells. However, the molecules or cells under study are frequently purified or freed from their 'normal' surrounding complex environments of the intracellular space or the extracellular matrix, respectively. It would be desirable to be able to apply optical trapping *in cyto* or *in vivo* for the study of such systems in a realistic environment. Furthermore, since cells, from bacteria to human cells, form part of larger 3D systems, it is clearly desirable to scale up experiments to multi-cellular systems, tissues or even whole organisms, particularly when considering signalling between cells, or physiological function. Some progress towards this has been made with, e.g., the trapping of otoliths ('ear stones') in zebrafish embryos, which can then be manipulated to stimulate a physiological response (Favre-Bulle et al. 2019). However, trapping, imaging and the all-important calibration remains challenging for a number of reasons: the trapping beam is distorted as it passes through the cell membrane; the interior of the cell is a crowded, complex, and spatially and temporally inhomogeneous environment; active processes in the cell (e.g. polymerisation and depolymerisation of actin) continually exert forces on a trapped particle, and so conventional methods based on predictable equilibrium thermal fluctuations cannot be used. These challenges will require scientific and technological innovation. This may come with the combination of optical tweezers with complementary techniques, such as super-resolution microscopy, spectroscopy or high throughput technologies.

A comprehensive review of the impact and possible future developments in optical tweezers in a wide range of fields can be found in the recently published Roadmap for Optical Tweezers (Volpe et al. 2022). We will leave the last word on future innovations, however, to the creator of optical tweezers, Arthur Ashkin (1999):

> As Yogi Berra said, "It's hard making predictions, especially about the future."

I predict, however, that optical trapping and manipulation of particles will continue to grow in those areas where they have unique advantages. Applications to atomic physics are proliferating and Bose-Einstein condensation and atom lasers are becoming separate fields of their own. New applications in the area of small macroscopic particle manipulation will continue to appear in areas such as studies of the forces between colloidal particles, colloidal crystallisation, entropic forces, and droplet formation and crystallisation in cloud physics. Applications to biology will continue to grow explosively in areas such as measurements on single mechanoenzymes, studies of living single cells, and organelles within cells (chromosomes, mitochondria, and nuclei). Finally, I predict the future will bring other new things, which you and I haven't thought of yet.

Philip H. Jones

References

Agrawal, R., Sherwood, J., Chhablani, J., Ricchariya, A., Kim, S., Jones, P. H., Balabani, S. & Shima, D. (2016), 'Red blood cells in retinal vascular disorders', *Blood Cells, Molecules, and Diseases* **56**, 53–61.

Agrawal, R., Smart, T., Norbe-Cardoso, J., Richards, C., Bhatnagar, R., Tufail, A., Shia, D., Jones, P. H. & Pavesio, C. (2016), 'Assessment of red blood cell deformability in type 2 diabetes mellitus and diabetic retinopathy by dual optical tweezers stretching technique', *Scientific Reports* **6**, 15873.

Ashkin, A. (1970), 'Atomic-beam deflection by resonance-radiation pressure', *Physical Review Letters* **25**, 1321–1324.

Ashkin, A. (1980), 'Applications of laser radiation pressure', *Science* **210**, 1081–1088.

Ashkin, A. (1992), 'Forces of a single-beam gradient laser trap on a dielectric sphere in the ray optics regime', *Biophysical journal* **61**, 569–582.

Ashkin, A. (1997), 'Optical trapping and manipulation of neutral particles using lasers', *Proceedings of the National Academy of Sciences* **94**, 4853–4860.

Ashkin, A. (1999), 'Optical trapping and manipulation of neutral particles using lasers', *Optics and Photonics News* **10**, 41.

Ashkin, A. (2000), 'History of optical trapping and manipulation of small-neutral particle, atoms, and molecules', *IEEE Journal of Selected Topics in Quantum Electronics* **6**, 841–856.

Ashkin, A., Boyd, G. D. & Dziedzic, J. M. (1963), 'Observation of continuous optical harmonic generation with gas masers', *Physical Review Letters* **11**, 14–17.

Ashkin, A., Dziedzic, G. D., Smith, J. I., Ballman, R. G., Levinstein, A. A., Levinstein, J. J. & Nassau, K. (1966), 'Optically-induced refractive index inhomogeneities in $LiNbO_3$ and $LiTaO_3$', *Applied Physics Letters* **9**, 72–74.

Ashkin, A. & Dziedzic, J. M. (1971), 'Optical levitation by radiation pressure', *Applied Physics Letters* **19**, 238–235.

Ashkin, A. & Dziedzic, J. M. (1989), 'Optical trapping and manipulation of single living cells using infra-red laser beams', *Berichte der Bunsengesellschaft für physikalische Chemie* **93**, 254–260.

Ashkin, A., Dziedzic, J. M., Bjorkholm, J. E. & Chu, S. (1986), 'Observation of a single-beam gradient force optical trap for dielectric particles', *Optics letters* **11**, 288–290.

Ashkin, A., Dziedzic, J. M. & Yamane, T. (1987*a*), 'Optical trapping and manipulation of single cells using infrared laser beams', *Nature* **330**, 769–771.

Ashkin, A., Dziedzic, J. M. & Yamane, T. (1987*b*), 'Optical trapping and manipulation of single cells using infrared laser beams', *Nature* **330**, 769–771.

Ashkin, A. & Gordon, J. P. (1979), 'Cooling and trapping of atoms by resonance radiation pressure', *Optics letters* **4**, 161–163.

Baresch, D., Thomas, J. L. & Marchiano, R. (2016), 'Observation of a single-beam gradient force acoustical trap for elastic particles: acoustical tweezers', *Physical Review Letters* **116**, 024301.

Belly, H. D., Stubb, A., Yanagida, A., Labouesse, C., Jones, P. H., Paluch, E. K. & Chalut, K. J. (2021), 'Membrane tension gates ERK-mediated regulation of pluripotent cell fate', *Cell Stem Cell* **28**, 273–284.

Bolognesi, G., Friddin, M. S., Salehi-Reyhani, A., Barlow, N. E., Brooks, N. J., Ces, O. & Elani, Y. (2018), 'Sculpting and fusing biomimetic vesicle networks using optical tweezers', *Nature communications* **9**, 1–11.

Borghese, F., Denti, P. & Saija, R. (2007), *Scattering from model nonspherical particles: theory and applications to environmental physics*, Springer Science & Business Media, Germany.

Chu, S., Bjorkholm, J. E., Ashkin, A. & Cable, A. (1986), 'Experimental observation of optically trapped atoms', *Physical Review Letters* **57**, 314–317.

Deufel, C., Forth, S., Simmons, C. R., Dejgosha, S. & Wang, M. D. (2007), 'Nanofabricated quartz cylinders for angular trapping: DNA supercoiling torque detection', *Nature methods* **4**, 223–225.

Philip H. Jones

Donato, M. G., Messina, E., Foti, A., Smart, T. J., Jones, P. H., Iatì, M. A., Saija, R., Gucciardi, P. G. & Maragò, O. M. (2018), 'Optical trapping and optical force positioning of two-dimensional materials', *Nanoscale* **10**, 1245–1255.

Essiambre, R. J. (2021), 'Arthur Ashkin: Father of the optical tweezers', *Proceedings of the National Academy of Sciences* **118**, e2026827118.

Favre-Bulle, I. A., Stilgoe, A. B., Scott, E. K. & Rubinsztein-Dunlop, H. (2019), 'Optical trapping in vivo: theory, practice, and applications', *Nanophotonics* **8**, 1023–1040.

Gauthier, C. R. (2005), 'Computation of the optical trapping force using an FDTD based technique', *Optics Express* **13**, 3707–3718.

Gieseler, J., Gomez-Solano, J. R., Magazzù, A., Castillo, I. P., García, L. P., Gironella-Torrent, M., Viader-Godoy, X., Ritort, F., Pesce, G., Arzola, A. V., Volke-Sepúlveda, K. & Volpe, G. (2021), 'Optical tweezers – from calibration to applications: a tutorial', *Advances in Optics and Photonics* **13**, 74–241.

Gore, J., Z. Bryant, Z., Nöllmann, M., Le, M. U., Cozzarelli, N. R. & Bustamante, C. (2006), 'DNA overwinds when stretched', *Nature* **442**, 836–839.

Han, X., Luo, H., Xiao, G. & Jones, P. H. (2016), 'Optically bound colloidal lattices in evanescent optical fields', *Optics Letters* **41**, 4935–4938.

Inanc, M. T., Demirkan, I., Ceylan, C., Ozkan, A., Gundogdu, O., Goreke, U., Gurkan, U. A. & Unlu, M. B. (2021), 'Quantifying the influences of radiation therapy on deformability of human red blood cells by dual-beam optical tweezers', *RSC advances* **11**, 15519–15527.

Jones, P., Maragò, O. & Volpe, G. (2015), *Optical Tweezers*, Cambridge University Press Cambridge, Cambridge, U.K.

Juniper, M. P. N., Straube, A. V., Besseling, R., Aarts, D. G. A. L. & Dullens, R. (2015), 'Microscopic dynamics of synchronization in driven colloids', *Nature communications* **6**, 1–7.

Juniper, M. P. N., Zimmermann, U., Straube, A. V., Besseling, R., Aarts, D. G. A. L., Löwen, H. & Dullens, R. P. A. (2017), 'Dynamic mode locking in a driven colloidal system: experiments and theory', *New Journal of Physics* **19**, 013010.

Kapitza, P. L. (1951), 'Dynamic stability of a pendulum with an oscillating point of suspension', *Journal of experimental and theoretical physics* **21**, 588–597.

Kelly, M. J. (1950), 'The Bell Telephone Laboratories', *Nature* **166**, 47–49.

Lebedew, P. (1901), 'Untersuchungen über die druckkräfte des lichtes', *Annalen der Physik* **311**, 433–458.

Loudon, R. & Baxter, C. (2012), 'Contributions of John Henry Poynting to the understanding of radiation pressure', *Proceedings of the Royal Society A: Mathematical, Physical and Engineering Sciences* **468**, 1825–1838.

Maier, S. A. et al. (2007), *Plasmonics: fundamentals and applications*, Vol. 1, Springer.

Maragò, O. M., Bonaccorso, F., Saija, R., Privitera, G., Gucciardi, P. G., Iatì, M. A. et al. (2010), 'Brownian motion of graphene', *ACS nano* **4**, 7515–7523.

Maragò, O. M., Gucciardi, P. G., Bonaccorso, F., Calogero, G., Scardaci, V., Rozhin, A. G., Ferrari, A. C., Jones, P. H., Saija, R., Borghese, F. et al. (2008), 'Optical trapping of carbon nanotubes', *Physica E: Low-dimensional Systems and Nanostructures* **40**, 234–2351.

Maragò, O. M., Jones, P. H., Bonaccorso, F., Scardaci, V., Gucciardi, P. G., Rozhin, A. G. & Ferrari, A. C. (2008), 'Femtonewton force sensing with optically trapped nanotubes', *Nano letters* **8**, 3211–3216.

Marko, J. F. & Siggia, E. D. (1995), 'Stretching DNA', *Macromolecules* **28**, 8759–8770.

Maxwell, J. C. (1873), *A treatise on electricity and magnetism*, Vol. 1, Oxford Clarendon Press.

Memoli, G., Fury, C. R., Baxter, K. O., Gélat, P. N. & Jones, P. H. (2017), 'Acoustic force measurements on polymer-coated microbubbles in a microfluidic device', *The Journal of the Acoustical Society of America* **141**(5), 3364–3378.

Mie, G. (1908), 'A contribution to the optics of turbid media, especially colloidal metallic suspensions', *Ann. Phys* **25**, 377–445.

Nichols, E. F. & Hull, G. F. (1901), 'A preliminary communication on the pressure of heat and light radiation', *Physical Review (Series I)* **13**, 307–320.

Pauzauskie, P., Radenovic, A., Trepagnier, E., Shroff, H., Yang, P. & Liphardt, J. (2006), 'Optical trapping and integration of semiconductor nanowire assemblies in water', *Nature materials* **5**, 97–101.

Pesce, G., Jones, P. H., Maragò, O. M. & Volpe, G. (2020), 'Optical tweezers: theory and practice', *The European Physical Journal Plus* **135**, 1–38.

Pesce, G., Volpe, G., Maragó, O. M., Jones, P. H., Gigan, S., Sasso, A. & Volpe, G. (2015), 'Step-by-step guide to the realization of advanced optical tweezers', *JOSA B* **32**, B84–B98.

Polimeno, P., Magazzu, A., Iatì, M. A., Patti, F., Saija, R., Boschi, C. D. E., Donato, M. G., Gucciardi, P. G., Jones, P. H., Volpe, G. et al. (2018), 'Optical tweezers and their applications', *Journal of Quantitative Spectroscopy and Radiative Transfer* **218**, 131–150.

Porta, A. L. & Wang, M. D. (2004), 'Optical torque wrench: angular trapping, rotation, and torque detection of quartz microparticles', *Physical review letters* **92**, 190801.

Poynting, J. H. (1903), 'The pressure of light', *The Inquirer* (Newspaper), 195–196.

Purcell, E. M. (1977), 'Life at low Reynolds number', *American Journal of Physics* **45**, 3–11.

Purcell, E. M. & Pennypacker, C. R. (1973), 'Scattering and absorption of light by nonspherical dielectric grains', *The Astrophysical Journal* **186**, 705–714.

Raab, E. L., Prentiss, M., Cable, A., Chu, S. & Pritchard, D. E. (1987), 'Trapping of neutral sodium atoms with radiation pressure', *Physical review letters* **59**, 2631–2634.

Richards, C. J., Smart, T. J., Jones, P. H. & Cubero, D. (2018), 'A microscopic Kapitza pendulum', *Scientific Reports* **8**, 1–10.

Smith, S. B., Cui, Y. & Bustamante, C. (1996), 'Overstretching B-DNA: the elastic response of individual double-stranded and single-stranded DNA molecules', *Science* **271**, 795–799.

Smith, S., Finzi, L. & Bustamante, C. (1992), 'Direct mechanical measurements of the elasticity of single DNA molecules by using magnetic beads', *Science* **258**, 1122–1126.

Stephenson, A. (1908), 'On induced stability', *The London, Edinburgh, and Dublin Philosophical Magazine and Journal of Science* **15**, 233–236.

Suresh, S., Spatz, J., Mills, J. P., Micoulet, A., Dao, M., Lim, C. T., Beil, M. & Seufferlein, T. (2005), 'Connections between single-cell biomechanics and human disease states: gastrointestinal cancer and malaria', *Acta Biomaterialia* **1**, 15–30.

Svoboda, K. & Block, S. M. (1994), 'Optical trapping of metallic Rayleigh particles', *Optics Letters* **19**, 930–932.

Uhlenbeck, G. E. & Ornstein, L. S. (1930), 'On the theory of the Brownian motion', *Physical review* **36**, 823–841.

Volpe, G., Maragò, O. M., Rubinzstein-Dunlop, H., Pesce, G., B.Stilgoe, A., Volpe, G., Tkachenko, G., Truong, V. G., Chormaic, S. N., Kalantarifard, F. et al. (2022), 'Roadmap for optical tweezers', *arXiv preprint arXiv:2206.13789* .

Wang, M. D., Yin, H., Landick, R., Gelles, J. & Block, S. M. (1997), 'Stretching DNA with optical tweezers', *Biophysical journal* **72**, 1335–1346.

Wu, M. C. & Ming, C. (2011), 'Optoelectronic tweezers', *Nature Photonics* **5**, 322–324.

Yan, Z., Sweet, J., Jureller, J., Guffey, M., Pelton, M. & Scherer, N. (2012), 'Controlling position and orientation of single silver nanowires on a surface using structured optical fields', *ACS nano* **6**, 8144–8155.

6 Are we alone in the universe?

ARIK KERSHENBAUM
University of Cambridge

Abstract: Is there intelligent life elsewhere in the universe? If not, does that mean that we humans are utterly alone in creation? Recent technological developments make the discovery of life on other planets almost expected within the coming decades. But most of the inhabited planets we hope to discover may well be populated by no more than alien bacteria. Will that make us feel any less alone? What we really hope to find are aliens with whom we can communicate and hold a conversation. When we ask ,"Are we alone?", what we really mean is, "Do we have anyone to talk to in the universe?" Our gnawing concern about being isolated in the universe ironically mirrors the situation we face on our own planet. We are proud of the status of human beings as the most intelligent of animals, and indeed the only species with language. But that very uniqueness isolates us from all the other intelligent animals on the planet. Sure, we can communicate with our pet dogs and cats, but we can't hold a conversation with them. Why not? What is the nature of the barrier between us and dolphins or chimpanzees? Some would say that if we aren't capable of understanding dolphins and whales, we have no chance of understanding any alien civilisation we encounter. Perhaps we are doomed to galactic isolation, no matter how many alien civilisations exist. However, I believe that we can be more optimistic than that. As we reach out to the stars to seek out new life and new civilisations, now is the time to consider: for what are we actually searching?

Introduction

"Are we alone?" It's the anguished question so often expressed not just by those who search for extraterrestrial life, but also by those who wonder about extraterrestrial life. And even by those who merely worry about the existence or absence of extraterrestrial life. Whether or not we are alone in the universe, isolated on a lonely rock, sailing endlessly and pointlessly through empty space, or whether we are one amongst billions of inhabited worlds, is a question that occurs to everyone at some point. Being the only intelligent species in all of creation seems like a terrifyingly lonely prospect to some, but a reassuring affirmation of our uniqueness and our special status in the image of God to others. Throughout all human history – ever since humans have had the consciousness to ponder existence – the answer to this question has remained elusive. In the past, our ancestors were convinced that higher spheres were occupied by others: angels, spirits, and demons. More recently, the rationalist revolution has swept away such imaginative ideas and replaced them with a simple question: if we are not alone, why have we not seen evidence of anyone? For the last century, science has provided little in the way of satisfying answers to our apparent isolation in the universe, and so, most people thought, we really are alone.

But we live in a unique and exciting time. The question of our isolation, or not, may be answered, decisively and conclusively, sooner than we think. We may be on the verge of finding other life in the universe. Just over 20 years ago, we weren't even sure whether planets even existed outside our solar system. Of course, it seemed likely that they did, given what we know about the way that stars are formed, and the likelihood that both rocky and gaseous planets would condense around a new-born star. But we had no quantitative feel for how many planets there might be, and how many of them might be suitable for sustaining life: Earth-like, with a rocky surface, as opposed to inhospitable gas giants like Jupiter. How astonishing that in those 20 short years, having examined more than 3,600 different stars, we have now confirmed the existence of over 5,000 planets. Given that our instruments are improving all the time, it seems likely that most stars have planets of some sort orbiting them. And a recent study estimated that there may be as many as 40 billion Earth-like planets in our galaxy (Petigura et al. 2013). Whether or not we are unique as humans, it certainly seems the case that our planet is not alone.

However, we are doing more than just finding interesting planets. New instruments like the James Webb Space Telescope are capable of peering into the atmospheres of these distant worlds and looking for spectrographic signatures of the chemicals that are present. Water would be interesting, of course, because all life we know of uses water, but possibly we could also detect oxygen, or other compounds that would seem to be convincing indicators of the presence of life. Oxygen is widely used by living organisms on our planet, but it isn't necessarily the foundation of life everywhere in the universe (indeed, early life on Earth didn't breathe oxygen at all). However, oxygen is such a reactive chemical that we would usually expect it to disappear very quickly from a planet's atmosphere, reacting with the rocks, for instance, and getting tied up as "rust" (this is why Mars is red). Finding oxygen in the atmosphere of a planet orbiting another star would be so unusual that it would need unusual mechanisms to explain its continued presence. It is such "unexpected" chemicals for which astrobiologists are searching, because unexpected chemicals imply unexpected mechanisms, and life is one such unexpected mechanism. Many scientists are now confident that convincing, if not conclusive, evidence of life-like processes on other planets will be forthcoming within the next decades, if not years.

So, what would be the implications of discovering life on other planets? Would we sense an end to our perceived isolation in the universe? Would it change our opinions of ourselves, our role and significance?

Well, probably not. Optimistic as scientists are about finding life in the universe within the coming years, that optimism is directed at the possibility of finding very simple life. Not life "like us". Whatever we find, it's unlikely to be something that will challenge our status as the self-declared preeminent creatures of creation. Life has existed on Earth for about 3800 million years, but for the first 3000 million years, there was nothing around that we would recognise as being at all similar to the complex animal and plant life that we see today. If life on other planets has evolved according to a similar history to that on Earth, then observing them at a random point in their history would most likely mean observing a planet covered in little other than bacterial slime. It's true that Earth is a relatively young planet, and it's quite possible that complex life has already had time to evolve elsewhere, but our atmospheric probing of other worlds may not be discriminating enough to distinguish between pond scum, and herds of wildebeest. In the absence of an alien David Attenborough, beaming nature documentaries to us from another planet, we are unlikely to know the precise details of what alien life is like.

It's a surprising suggestion, but I would claim that discovering bacteria on another planet is unlikely to move, inspire, or disturb most people. Yes, it would cause an earthquake among scientists, but how would this discovery affect ordinary people? It's true, the inhabitants of our planet will no longer be able to consider ourselves to be the unique forms of life in the universe. That status will have been removed from us, but we will remain as isolated as ever. Alien bacteria will not change much about the course of life on Earth – after all, alien bacteria won't particularly want to make friends with us. Naturally, if we were to discover an advanced civilisation, capable of communicating with us over the vast interstellar distances, our perspective on our cosmic isolation would change. But this is not only unlikely to happen, we also have no metric for estimating when it might happen. I can say with some confidence that I estimate we will discover signs of life on other planets, because I can extrapolate from our discoveries to date, taking into account the planned improvements in sensing technology. But I cannot make a reasoned estimate of the likelihood of finding advanced technological civilisations, without any data on which to base such a claim. Aliens might arrive tomorrow and welcome us into the galactic community. Or they might never come at all. We have no way of knowing. So, while the answer to the question, "Are we alone?" may very well be "No", nonetheless, if we encounter aliens with whom we can't communicate, we will probably still feel just as isolated as ever.

Not being alone

What fascinates me about being not alone (i.e., not being the only lifeforms in the universe), and yet being isolated nonetheless, is that such a galactic scenario mirrors very closely the situation on our own planet. Are we alone on Earth? Of course not. We are surrounded by life, both complex and simple. When you take a walk in the woods, you might describe yourself as being "alone", or "isolated", but deep down you don't feel anything of the sort. You are aware of the trees and the birds, and you interact with them: you respond emotionally to the songs of the birds, and you marvel at the majesty of the trees. What's more, there is some life on Earth with whom we have a very close bond; life that quite directly and explicitly provides us with attachment – our pets, for one thing. Why is it that having no one else to talk to in the universe makes us alone, whereas we do not feel alone with dogs, cats, birds, and trees? The question of whether or not we are alone, either in the universe, or in our homes, may be a little more complex than it seems at first.

Why, then, do we pose the question "Are we alone?", when we don't really seem to be asking that question at all? Why do we feel that we would be excited to discover life on another planet, but we are not amazed and awed by the dog who keeps us company at home? What are we actually looking for? Will only humanoid aliens provide us with an acceptable alternative to galactic isolation?

I have worked in many isolated environments over the years: in the jungles of Vietnam, the vast expanse of the Rocky Mountains, the desolate

Arik Kershenbaum

Kalahari Desert. Many years ago, I made my first trip to the Serengeti in Tanzania. There, without a human being in sight, I was surrounded by life. Zebras and lions, impala and leopards. And I was overwhelmed by a feeling of irrelevance – that my existence as a human being was of no importance whatsoever to the zebras and the lions and the impala. The whole ecosystem would function perfectly well without me, and I was unimportant to them. There was a barrier, isolating me from the rest of the teeming life of the plains. Something is different about us, and them, and that difference serves to isolate us from other life on Earth. We humans are isolated on our own planet – because we cannot talk to anyone else.

Recasting the question "Are we alone?" through the lens of language gives us a new and useful perspective. It seems, perhaps, that we won't be satisfied with aliens unless we can communicate with them. When we ask "Are we alone?", what we really mean is, "Do we have anyone to talk to in the universe?" Similarly, we feel that we are unique among life on our planet, unique and isolated, despite being surrounded by life, because we cannot speak with them. So, discovering intelligent aliens – that is, aliens who can communicate with us in a way we would understand – really would be an extraordinary event, which would elevate us from a status of "alone", both on our planet, and in our galaxy, to "not alone".

However, although we do not seem to be able to "talk" to our fellow non-human inhabitants of planet Earth, it certainly seems to be the case that we can communicate with them in some way. At the very least, we have an intimate relationship with our pets, and a certain amount of mutually under-stood two-way communication is commonplace. "Do you want to go for a walk?" is misunderstood by no dog ever; and nor is their response of ex-cited tail wagging and hovering next to the front door. Still, despite such a deep emotional bond, there nonetheless appears to be something missing from our relationships with non-humans, something that convinces us that we are, in practice, alone on our planet. We view a dog as our "best friend", and we share with our dogs 3700 out of 3800 million years of our evolu-tion, but we cannot really share our deepest secrets and desires with them. With an intelligent alien, we share no biological or evolutionary connection whatsoever, and yet they might be capable of discussing with us science, literature and politics. We would feel less alone with a 12-legged talking alien than we do with a chimpanzee, with whom we shared a family ances-tor just six million years ago. Why can't we talk to animals? We certainly want to, and our stories and legends are full of references to people who can talk to animals, and animals that can talk like people. It seems to be important to us – and yet doesn't seem practical. Why not? One possibility, of course, is that we are just not smart enough: we haven't tried hard enough. Some scientists are still trying to "decode" animal communication, and believe that if we work hard enough, one day we will be able to hold a conversation with dolphins, chimpanzees, and parrots. Another possibility is that the animals themselves are not smart enough. That we humans have evolved some essential trait – an essence of language – that has not evolved in any other lineage. If so, then any attempts to teach chimps English, or to whistle conversations with dolphins, is doomed to failure.

What is life?

Astrobiologists often ask the question, "What is life?" This is, indeed, an important question when searching those distant planets for signs of "life" – whatever that is. If we can't define life, it's hard to know how to search for it. However (without detracting from the importance of searching for such a definition) it seems likely that we will know life when we see it. Life is inherently out of equilibrium with the rest of the universe: we live, grow, and reproduce, whereas most of the universe simply does not behave in that way. We also accumulate complexity through natural selection – really the only way that life-like complexity can accumulate. If we were to observe such self-sustaining and evolving systems, we'd certainly find them exceptionally interesting, even if we couldn't quite agree on whether or not they are alive. I don't believe that the struggle for a definition of life is a particularly debilitating struggle in astrobiology.

Not so language. We haven't yet discovered bizarre life in the universe over which to puzzle "is it alive?", but we are and always have been surrounded by communicating animals, and we still struggle to decide whether or not they have a language, and whether their communication is intentional or not. We have an intuitive feeling that there is a clear difference between something that is alive and something that is not alive, whereas we have no intuition at all about what is talking to us and what isn't. This crucial difference is important because, so it seems, it's not alien life in itself that determines whether or not we are alone, it is talking alien life. Emerging from our isolation requires us to recognise extraterrestrial interlocutors, in a way that we haven't so far managed to identify any terrestrial interlocutors.

Arik Kershenbaum

Language

Can we devise a test to determine whether a species – alien or terrestrial – is talking to us? Such a test would be useful in our exploration of the universe, and in our attempts to find intelligent and technological alien civilisations. If we receive a mysterious radio signal from outer space, we need an algorithm to determine whether or not that signal was intentional, and whether or not it contains actual information. Such a test would have to be extremely general. Computational linguists on Earth have devised a series of mathematical descriptions of language, which appear to conform well to all the varieties of our human language but may nonetheless be biased by our own experience – we've never heard any language other than human. The design of a test for a "language fingerprint" is a crucial part of the search for extraterrestrial intelligence. But if we could devise a general and universal test for language, what would we find when we applied that test to animals on Earth? Would we discover that those birds on your walk through the woods are actually talking about you behind your back? Would it turn out that dolphins are really saying "so long and thanks for all the fish"?

In our search for a good definition and test for language we must be careful not to bias ourselves into a narrow definition of language that automatically excludes every species on Earth except for humans. But at the same time, we need to recognise that what we call "language" is something distinct, something qualitatively different from "not language". If we lose that distinction, if we admit all forms of complex animal communication into the umbrella of language, we have performed ourselves a disservice. If all communicating creatures have language, then we cannot usefully answer the question, "Are we alone?", except to say, "We are never alone, as long as we have the dogs and the dolphins and the beetles around us". And, by extension, the alien bacteria. Language is something that is not necessarily unique to humans, but is certainly something rare, and if it is shared with other species, it is shared with very few. So, it is helpful to look at the nature of human language, and in particular, the evolution of human language, so that we can find a suitably universal, and yet restrictive definition.

Why evolve language?

Why did humans evolve language? And do we dare postulate that any alien species that has language will have evolved that language along a similar evolutionary path to our own? Of course, these are very difficult questions to answer. Language leaves no fossils, and so we have no empirical record of the slow steps, generation by generation, from our ancestors who had no language, to those who first could be said to possess that trait. The evolutionary transition from non-language to language will probably forever be unknowable. But as evolution tends to make traits become more useful, more beneficial to a species, we can look at the possible reasons why language might have been advantageous to our ancestors and consider whether these reasons might be general enough to provide a universal test for language.

In fact, there would appear to be a vital connection between language and the nature of isolation. It goes without saying that communication itself, from the roaring of tigers to the chemical pheromones of moths, only exists both because of isolation, and because of a lack of isolation. If there were no isolation between animals, there would be no need to communicate. And if animals were completely isolated, there would be no advantage to communication. Complete integration between individuals means that everyone would know immediately what the other were thinking, and for such sci-fi drones (it's got to be science fiction, because complete integration never really exists) there would be no advantage to evolving any form of complex communication like language. On the other hand, complete isolation between individuals would mean that such hermits would be wasting their time attempting to communicate with others. The evolution of complex communication – and hence of language – requires a balance between isolation and integration.

Animals interact for many reasons – most notably to find a mate, but also to form groups for protection from predators, or to achieve communal tasks, like finding food or keeping warm. In those animals where social interactions exist, information must flow between individuals. The animals themselves cannot survive without their social groups, and so they cannot survive without communication. It is precisely the lack of isolation – combined with the danger of isolation – that causes complex communication to evolve. Communication, therefore, is ubiquitous in animals, and even in many plants. What about language, though? Why is it that all animals communicate, but only humans (it would seem) have a language?

One theory would suggest that our ancestors evolved language because complex communication was necessary to support the kinds of functions that our ancestors needed to perform. If we needed to hunt a mammoth, we needed to communicate to our peers the hunting strategy and the plans for how to execute that strategy. However, such a theory lacks evolutionary rigour. Sure, language would be useful in coordinating hunting, if we already hunted in a coordinated way – but how could we coordinate hunting without language in the first place?[1] There is a chicken-and-egg paradox here: which came first – the ability to use language, or the need to use it?

Rather more convincing is the idea that human language evolved as a result of increasing complexity in our brains, irrespective of the adaptive benefit that language would provide in the future. Our ancestors lived in large groups, it is true, but groups not much larger than chimpanzees, or dolphins, and certainly not larger than wildebeest. What made the difference was the nature of the interactions between individuals in those social groups. The individuals were integrated – they interacted and performed the functions of the society in a coordinated way – but they were at the same time isolated in their own minds. Our ancestors had, and we still have now, our own internal mental states, our own thoughts, our desires and goals, and importantly, our plans. We manipulate, we strategise, and we take advantage of others, by knowing both what we want and also what others want. It may well be that language evolves only when individuals have both a need to communicate complex ideas, and brains that are capable of understanding that others have complex ideas too – complex ideas that may very well conflict with my own. That balance between integration and isolation seems to fit with what we understand about the societies of our near-human ancestors, and also fits with the lack of language in other Earth species like dolphins and chimpanzees. Could such a criterion also apply to the evolution of language on other planets?

I find evolutionary explanations of the fundamental nature of language appealing. We don't know anything about the physical nature of life on other planets, or about the ecosystems in which such life exists. But we do have every reason to believe that the laws of evolution are universal and are indifferent to the biochemical specifics of weird and wonderful alien life. Whether life on other planets is based on a carbon chemistry or not, whether aliens have four limbs or 24, whether they live on the surface of a rocky planet or in an underground ocean; evolution will proceed according to the rules we know and understand from examining life on Earth. And this applies to the way that animals, or their alien equivalent, interact with each other and communicate with each other. I don't know whether alien

---

[1] The reader might point out that many animals without language hunt cooperatively – wolves, for example. However, while cooperation in nature is common, coordination is extremely rare. Wolves do not coordinate their roles while hunting.

language will have nouns and verbs, or even whether such a language will consist of words at all. But I know it will arise from a need to communicate some things, and at the same time, a need to keep some things secret. Only in such a society of organisms would language provide any evolutionary advantage.

Teeming with life, probably

We don't actually know if we're isolated in the universe. For now, it would appear so. The universe is probably teeming with life, but that doesn't necessarily make us any less alone. If there are creatures like us waiting out there, then at least we know in advance that we share something in common. The need to communicate, and the ability to communicate in a way that seems to us like language, would make all the difference between being "alone" in the universe (in the way we might be "alone" in a forest) and being part of a galactic community of talking beings – even if we haven't met any of them yet.

Not for nothing are humans sending messages into space, looking for creatures who communicate like us. When we design signals that we hope to be intelligible by other civilisations, we work on the basis that such a civilisation will share certain things with our own. If that were not the case, we would probably never be able to communicate with them anyway. Any alien civilisation we succeed in contacting must have a technology of some sort – otherwise they could hardly build the equipment necessary to receive our signals – and to develop such a technology they must be social and cooperative creatures, and to be social and cooperative: well, perhaps that means they must have language too. We may not find any interlocutors in our lifetime, but at the very least we know what we are looking for.

Will we even be capable of interpreting any message from outer space, capable of understanding and decoding what aliens might have to say? This is a challenging question, and one on which scientific opinion is split. Perhaps it is simply not possible to decode, or even attempt to decode an utterly alien language, without the ability to hold a genuine two-way conversation with the other side. Linguists say that dialogue itself is the nature of language, and the vast distances between the stars means that it might be decades, or even centuries before we receive a reply to any question we ask. Isolation by communication is actually exacerbated by physical isolation

Some people would say that if we can't understand what dolphins and

Arik Kershenbaum

birds and wolves are saying, then what chance do we have to understand aliens? We might be surrounded by others like us, but we will be forever isolated, forever locked in our own bubble as unintelligible foreigners in a galaxy of civilisations speaking languages we will never understand. But let's not be disheartened by our inability to understand animals on Earth. In the end, they do not have a language of their own. No matter how tightly bonded to our pet dogs and cats we are, no matter how much we can understand their desires and moods, they are still far more alien to us than our potential friends on other planets. Our shared evolutionary history with dogs ended 95 million years ago, and our ancestors went our separate ways. In contrast, we are utterly isolated evolutionarily from any alien civilisation, without so much as a shared hour of evolutionary history. Nonetheless, those aliens – if they have a civilisation of their own – likely have traversed similar challenges and found similar solutions to our own ancestors. We share a lot more with those new friends waiting for us than we think.

References

Petigura, E. A., Howard, A. W. & March, G. W. (2013), 'Prevalence of earth-size planets orbiting sun-like stars', *Proceedings of the National Academy of Sciences* **110**, 19273–19278.

7 Self-imposed isolation of North Korea

HEONIK KWON
University of Cambridge

Abstract: North Korea is one of the most secluded societies in today's world. Its system of rule is often referred to as an enigma of modern politics. This essay asks what has caused this condition of extreme isolation, highlighting the relentless pursuit of a historically durable charismatic political power. The discussion will include Max Weber's thoughts on the place of charismatic power in modern politics.

Introduction

North Korea is commonly referred to as one of the world's most secluded and enigmatic places. Travel to and from the country is highly restricted, as is mobility within the country for its twenty-five million citizens. The country's state hierarchy is bent on shielding its society from the gaze of the outside world and keeping its population from discovering a way of life other than the one offered within the country. The country's leadership also has a strong interest in maintaining an enigmatic appearance among its people. The prevailing interest involves the state's preparedness to employ coercive measures against individuals and their families who happen to fall from the webs of mystical moral symbols and ideas that are spun around the historical integrity of the North Korean way of life.

On the diplomatic terrain, Pyongyang currently has very few friends and allies despite its maintenance of formal diplomatic ties with a number of states especially in the Global South. China is its closest and strongest ally, and North Korea's state media propagates blood-forged perpetual friendship with China tracing back to the time of the Korean War. However, this friendship is far from set in stone and has had ups and downs during the past seven decades. Many Chinese see their southern neighbour as a liability rather than a friend, and rather strong feelings against China are prevalent in the Workers' Party of North Korea, especially among its rank-and-file members.

None of this makes North Korea truly exceptional, as there are other political societies in today's world that are keen on controlling the flow of information or human mobility for political or religious reasons. North Korea is an exceptional case in the extent and intensity of its readiness to isolate itself from the international community when it deems it necessary to do so. It does not have an isolationist ideology and has a contrary historical background, as will be mentioned shortly. Today, however, North Korea is adopting strong rhetoric akin to that of extreme isolationism. For instance, Pyongyang recently unleashed slogans such as "Let us live in our own way!" and "Whatever they say, we go the way we choose to go!"[1]

In his 1981 essay titled "North Korean Enigma", Jon Halliday (1981) writes the following:

> North Korea, the Democratic People's Republic of Korea (DPRK), is an isolated enigma in Northeast Asia. No state in the world lives with such a gap between its self-image and self-presentation as a socialist 'paradise on earth' and the view of the most of the rest of the world that it is a bleak, backward work-house ruled by a megalomaniac tyrant, Kim Il Sung.

Political system

However, there is actually no mystery about the North Korean political system. The North Korean state is not an enigmatic entity and never has been. North Korea's founding leader, Kim Il Sung, was a skilful political actor who knew how to build an aura of captivating charismatic power. He understood the efficacy of this power for mobilising the masses towards ambitious political goals and he was committed to keeping that power not only during his life but also beyond the time of his rule. Modern world history abounds with similar charismatic, visionary leaders and stories of their rise and fall. The same is true in the political history of the communist world that constituted the moiety of the Cold War international order. This world was distinct from the other half of the global order of the era not only in

---

[1] *Rodong Sinmun,* 9 November 2018.

Heonik Kwon

terms of its mode of regulating economic activity but also in its ways of pursuing the modern ideal of a secular, disenchanted society free from traditional beliefs and backward ideas. We know that the disenchantment of society pursued in revolutionary socialist polities involved much more explicit and conscious intervention by state power than in liberal capitalist societies. However, we also know that the performance of secular revolutionary politics, while aiming to demystify traditional religious norms and mystical ideas, nevertheless often involved the mystification of the authority and power of the revolutionary leadership.

The evolution of North Korea's postcolonial political system has not been an exception to this well-trodden general trend of modern revolutionary politics. North Korean leaders imported foreign political ideas from other, more powerful states, particularly the Soviet Union, and transformed them according to their own aims, adding some creative indigenous elements and facades. North Korea's political genesis is fundamentally no different from the experience of many other newly independent postcolonial states of the twentieth century, which, while consolidating a political community by applying the established techniques of state and nation building borrowed from earlier European exemplars, typically claimed that the process was an exclusively indigenous, national art of politics. From this perspective, the North Korean political system is just as modern and as much a product of interaction with global modernity as any other political system existing in the world. In this respect, Bruce Cumings is right to claim that North Korea is nothing more and nothing less than "another country" in the modern world (Cummings 2004).

Although the character of the North Korean postcolonial political system is not unique in history, North Korea is unique in maintaining this particular character for longer than any other state entity born in the early Cold War and indeed far beyond the end of the Cold War as the prevailing geopolitical order of the twentieth century within which it took shape (initially by the Soviet occupying power) and evolved. The early North Korean political order was centred on an able and pre-eminent personality, as were the orders of other revolutionary states known in the history of the Cold War. This personality, Kim Il Sung, was in substance and form no more extraordinary than other leading twentieth-century revolutionaries – notably, Joseph Stalin and Mao Zedong. These leaders had prestigious careers in emancipatory political movements and each of them led a mass-based yet elite political organisation which, harbouring the principle of democratic centralism, was focused on mass mobilisation for radical social transformation. They grasped the central importance of modern technology in politics, including the effectiveness of print technology, art, theatre and drama, for mobilizing the masses. They also knew very well that an elite revolutionary vanguard organisation was not always an easy family to run, and that the efficient functioning of this organisation, at times, required an exceptional, charismatic leader whose authority transcended the realm of institutional politics.

The historical lives of these charismatic revolutionary leaders of the twentieth century can be discussed not only in terms of comparative history but also according to the conceptual premises of historical sociology – most notably, those of the eminent theoretician of modern politics, Max Weber. Weber was interested in the typology of modern political power and authority, including charismatic authority (Weber 1947). No matter how strange the phenomenon may appear to rational eyes, according to him, the enchanting power of charismatic authority is an intelligible historical and social phenomenon whose nature is no more mysterious than that of traditional authority (e.g., the authority of an emperor) or that of modern bureaucracy and legal systems. Weber understood that all these forms of human authority are imperfect, yet despite their imperfect nature, they all aspire to perfection and frequently claim to be perfect. When an apparently extraordinary charismatic authority appears on the horizon, according to him, the circumstances of its rise may be other than ordinary, but its constitutive order is nothing but extraordinary. In Weber's view, there is nothing miraculous about the miracle-claiming personality, and charismatic authority exists only because of the imperfection of other authorities. Charismatic personalities erupt in history in situations of radical social upheaval, when society's aspiration for change can no longer be contained within the routine traditional order or be satisfied by an existing legal-bureaucratic order. However, Weber also clarifies that the historicity of charismatic authority, because it originates in extraordinary times of social crisis, is limited in time and eventually dwindles as the society recovers from the upheaval and returns to a routine, everyday order. Most of the charismatic, cultic state personalities of the mid-century Cold War era underwent a dramatic rise and fall following the historical destiny of charismatic authority envisioned by Weber at the turn of the twentieth century – except in North Korea.

The exceptional character of the North Korean political system is, therefore, not the specific relationship between the state and the society, anchored in what we commonly call the cult of personality; rather it is the fact that this particular mode of rule has shown remarkable resilience, defying the contrary historical trend found in most other revolutionary and post-revolutionary societies. The durability of this form of politics is an exception in a theoretical sense as well, defying the historically impermanent nature of charismatic politics rendered in the Weberian exposition of modern political power and authority. The enigma of the North Korean political system is, therefore, not the practice of an extraordinary cult of personality but the extraordinary continuity of this practice. In contemporary North Korean political terminology, the country's unique, protracted and cross-generational charismatic politics is called, among other expressions, "legacy politics" (yuhun jŏngch'i).

My earlier work investigates the historical origins of North Korea's charismatic politics (Kwon 2010). In it, I question how the North Korean political system overcame the impermanent nature of charismatic authority and how it achieved what Weber calls the "routinization of revolutionary charisma" for such an astonishingly long period of time and in an apparently hereditary form of succession of rule. As recent events have shown, North Korea's legacy politics, now into the third generation, continues many years after the passing of the North Korean polity's founder in 1994. The first succession of power from Kim Il Sung to his eldest son, Kim Jong Il, was "the communist world's first hereditary transfer of power."[2] With the establishment of the third-generation hereditary leadership in 2011, the country undertook new projects to build up the authority and majesty of the paramount leader's persona and enable the continuity of the political order free from the risk of a rupture in the political life of the charismatic authority.

I argue that the evolution of North Korea's statehood has been an epic struggle against the impermanent nature of charismatic authority and against the mortality of this authority, to which all other charismatic personas of the twentieth century eventually succumbed, and that this struggle continues to date. Then, we can also consider the risks of a rupture – challenges to the durability of the paramount leadership that originate from the broad external environment. Most notable in this respect in recent times was the disintegration of the Soviet order and the Soviet-led order of international socialism in the early 1990s. The end of the Cold War was a huge shock to Pyongyang. The upheaval soon resulted in an unprecedented economic and human crisis, which involved the collapse of trade and industrial

---

[2] The quote is from "North Korea Confirms Kim Jong-Il's Son Will Take Over as Leader," The Guardian, 8 October 2010.

activity, a rapid decline in agricultural output and tragic failures in public health and food distribution. The magnitude of these failures was such that the entire system of North Korea's socio-economic production and distribution had become largely dysfunctional by the mid-1990s, and the combined effects of economic and administrative failures soon developed into a devastating famine and massive loss of human life. These catastrophic social crises coincided with the death of the country's revered founding leader, Kim Il Sung, which sank the entire nation into a deep spiritual crisis. A powerful ideology arose from this environment of generalised crisis which turned the telos of North Korea's party state from the caretaker of the people's economy to the defender of its political sovereignty. This ideology was called military politics or *sŏn'gun*, the power of which is manifested today, above all, as a claim to nuclear-armed statehood.

Painful isolation

The end of the Cold War thrust North Korea into painful isolation, and it took time to digest the meaning of the historical rupture. A conference was held in Pyongyang in August 2006 among North Korea's prominent political theorists to discuss the genesis and theoretical premise of the military-first political idea. The conference concluded that "the Dear Leader's [Kim Il Sung's first successor, Kim Jong Il] passionate ideological-theoretical activity advanced the Great Leader's [Kim Il Sung's] *sŏn'gun* ideology to a theoretical principle of revolutionary politics and to a great guiding principle for our times."[3] Other North Korean documents on the theory of military-first politics published in the 2000s claim Kim Il Sung's and Kim Jong Il's joint authorship of the theory. An authoritative text published in 2004, Understanding *Sŏn'gun Politics*, for instance, carefully manoeuvres between providing a genealogical explanation for the origin of *sŏn'gun* and a situational analysis of the idea's evolution. While locating the origin of military-first politics in the distant history of the anti-colonial resistance led by Kim Il Sung in 1930s Manchuria (against Japan's colonial rule) and armed struggle for national liberation in the early 1950s (the Korean War), it advocates that the idea took concrete form during the long interaction between the Great and Dear Leaders. It also claims that although the military-first political idea has a deep historical origin, its full theoretical force has been realised only in the Kim Jong Il era owing to the Dear Leader's vigor-

---

[3] *Rodong Shinmun*, 17 August 2006.

Heonik Kwon

ous intellectual efforts to protect North Korea's revolutionary heritage at a time of grave crisis in international socialism in the 1990s.

Understanding Sŏn'gun Politics starts, as many other contemporary North Korean social science publications do, with a quote from the Dear Leader: "Entering the last decade of the twentieth century, socialism collapsed in the Soviet Union and in countries of eastern Europe; this has resulted in great changes in global political structure and relations of power (Chun 2004, p.1)". A few pages later, the book introduces another quote from the Leader:

> Due to the imperialist reactionaries' plot against the Republic [Democratic People's Republic of Korea] that sought to isolate the Republic and to press it to death, our revolution came to confront cruel challenges and obstacles unprecedented in history. We became a lone fighter against American imperialism and against the concerted aggression from the imperialist forces (p.6 Chun 2004).

Based on these primary citations (featured in a bold script, distinct from the rest of the text, according to the established printing tradition in North Korea), the text presents a readable, interesting analysis of the implications of the collapse of international socialism for North Korea in the first part of the book. Later, the analysis moves on to the main conceptual premises of military-first politics, arguing that this politics, spearheaded by the Dear Leader's genius, was the only possible philosophical thesis and the best theory of the North Korean socialist revolution in the hostile global environment of the 1990s.

About the collapse of the international socialist political order in Russia and Europe and the consequent end of the Cold War as a geopolitical order, *Understanding Sŏn'gun Politics* has the following things to say. It argues that the bipolar world order of the Cold War era, after the end of the global conflict in 1989 to 1991, has not evolved into a peaceful, multipolar international order. Instead, the world has degenerated into a unipolar world order dominated by American power and rife with political conflicts and threats of war:

> Conflicts during the Cold War were mostly provoked and radicalised by the reciprocal enmity between the era's two superpowers, the Soviets and the Americans. After the end of the Cold War, by contrast, conflicts have arisen between nations, ethnic groups and political factions, and these were ignited by the contradictions resulting from the uncurbed preponderance of American power (p.3 Chun 2004).

An emboldened post-Cold War United States, and in the absence of coun-
tervailing forces provided by the international socialist order, according to
the text, concentrates its aggression against anti-imperialist forces in the
Third World seeking self-determination. Given this volatile condition that
poses a grave challenge to the global revolutionary forces, the text concludes that "after the breakdown of socialism in some countries and the
end of the Cold War, the global political order has changed from the formation centred on the Soviet-U.S confrontation to one that is based on the
contest of power between [North] Korea and the United States (p.15 Chun
2004)." The Workers' Party newspaper *Rodong Sinmun* asserts that owing
to the leadership's military-first politics, "Northeast Asia and the world at
large now benefit from having a new pole of justice [North Korea] that can
confront the power of super-empire in the unipolar world."[4]

These assertions may sound astonishingly self-centred and as a gross ex-
aggeration of the power of North Korea, yet they are based on the following
assessment of North Korea's place in the post-Cold War world order:

> The flag of socialism was taken down in the former Soviet Union and
> Eastern European countries. In the broader international sphere, peo-
> ple who long for socialism are thrown into confusion and left with
> no guidance. At this testing time, we refused to make any change.
> Instead, we raised our red flag of socialism even higher than be-
> fore. This way, our country became the only remaining bastion of
> socialism and was illuminated with the esteemed honour of doing so
> (p.7 Chun 2004).

At the same time, according to Understanding Sŏn'gun Politics, North Korea became the sole vanguard of the Third World revolution and a leader
among developing nations in their collective struggle against coerced in-
corporation into the "new world order" orchestrated by American imperial
power, and thus "the only source of light that can ignite the fire of self-
determination among peoples in the Third World" (pp.8-9 Chun 2004). As
the text argues, this singular, vanguard position of North Korea in the post-
Cold War world explains why the United States feels so threatened by the
existence of North Korea and, likewise, why U.S. news media (such as the
New York Times) call North Korea "the world's most dangerous country"
(p.11 Chun 2004).

---

[4] *Rodong Sinmun*, 24 August 2010.

A set of intriguing issues arises from the above rendering of contemporary world politics as the background of the rise of North Korea's military-first politics. Notable is the interpretation of the disintegration of Soviet political unity. *Understanding Sŏn'gun Politics* argues that North Korea took over the former Soviet Union's position as a main contender with American power in the post-Cold War world order. The substitution of North Korea for the Soviet Union as the leader of a global socialist revolution was not North Korea's wilful choice; rather, this was enforced upon it because it had become, in the new world order, the only existing revolutionary polity keeping intact the proud flag of socialism. The text discusses what it takes to achieve this, based on a strong assertion that the most fundamental reason for the disintegration of the Soviet order was the de-politicisation of the Soviet armed forces. In this light, *Understanding Sŏn'gun Politics* presents a highly negative view of the reform measures taken by the former Soviet leader Mikhail Gorbachev in the second half of the 1980s, which it argues were aimed at separating the Soviet army from Soviet politics. This view is in direct conflict with the conclusion reached by many Western observers about this period, who typically present a generally positive assessment of Gorbachev's role in ending the Cold War in Europe, citing particularly his refusal to intervene militarily in the political crises that the Soviet Union's European allies were undergoing at the end of the 1980s. Melvyn Leffler's authoritative history of the Cold War, for instance, assigns the most pivotal role in ending the long-drawn U.S.-Soviet conflict to Gorbachev's exemplary efforts towards denuclearization and his search for non-military solutions to international conflicts and crises in Eastern and Central Europe as well as in Afghanistan (Leffler 2007). The assessment of this critical period in *Understanding Sŏn'gun Politics* draws a very different conclusion, arguing that Gorbachev's military reform and his opting for a non-military solution to the crisis in international socialism resulted in the fall of the Soviet army from the status of the Party's army and the army's loss of its fundamental character as the army of socialism and the proletarian class, which opened the door for the dissolution of the Communist Party, the collapse of the Soviet rule, and the return of capitalism (p.44 Chun 2004). On the basis of this assessment, the text asserts the following:

> The tragic fate of the former Soviet power shows that the army cannot retain its class-based character if it is separated from the Party's leadership. It shows that the Party will fail its task and collapse itself unless it holds the army under its grip. From this analysis comes the principle that a victory in revolutionary struggle is viable only when

the Party of the proletarian class holds the army within its power and when the army devotes its strength solely to defending the party. The principle is according to the law of revolution that the Army is the Party and that the Party is the Army (p.45 Chun 2004).

Also notable is North Korea's self-definition as an alternative world leader after the collapse of the Soviet empire. The text advances this highly debatable, self-centred understanding of the contemporary global political order as centring on a North Korea versus the United States power nexus, drawing upon supporting evidence from selected Western sources. These sources include *The Voice of America*, which allegedly states, "North Korea is recognised as *the most influential basis*, in northeast Asia as well as in the broader global terrain, for the revival of socialism." (p.8 Chun 2004). This and other references to foreign sources within the text interact with extracts from the authoritative discourse of the Dear Leader. The former are surely meant to generate the impression that the argument presented in the text is a reasoned, scientific analysis based on knowledge of world opinions and trends. Like most other referential practices in contemporary knowledge production, the references to foreign sources introduced in this and other North Korean texts about military-first politics are also meant to strengthen the authority of the text's arguments. However, the ultimate authority for the argument, within the textual reality, clearly lies in the occasional, bold-scripted citations from the works and speeches of the country's leader rather than sources from the outside world. Concerning this particular referential strategy, it is worth adding that the meaning behind the text's allusion to the leader's authoritative voice and knowledge goes beyond what we would normally understand with the citational practice. In reading *Understanding Sŏn'gun Politics* as well as other analytical literature about military-first politics, it is apparent that the references to the leader's spoken or written words are not merely meant to draw authority from the latter in support of the presented argument but, more importantly, to augment the authority and esteem of the cited discourse. For example, the statement that North Korea is brought to confront American imperialism all alone, which the leader is quoted to have said about the destiny of North Korea after the collapse of the international socialist alliance, evolves in the narrative presented in the text to the somewhat astonishing conclusion mentioned above that North Korea has replaced the Soviet Union in the new post-Cold War global order as the only capable opponent of American power. Furthermore, the text employs the series of references to foreign sources for the specific purpose of justifying the argument, which is intended to magnify the stately status of North

Heonik Kwon

Korea and thereby the prestige of the country's leadership. The act of citing
what the leader said about the world, therefore, is aimed primarily at raising
the honour and authority of the speaker, which may involve amplifying and
even radicalizing the meanings of what the leader actually said.

Related to the idea that North Korea and the United States make up a
new bipolar world order after the collapse of the old bipolar geopolitical order, it is also interesting to observe that the text's analysis of the post-Cold
War, post-socialist world order focuses singularly on one particular stream
of global change as being against another equally important developmen-
tal stream. There have been two separate processes of post-socialist social
development: one of them mainly associated with the former Soviet Russia and Eastern Europe and the other with Asian socialist polities, notably
China. The latter is typically referred to as market socialism or socialism
with a market orientation and characterised as a mode of social development
combining a centralized political rule with controlled market economic de-
centralisation and liberalisation. This mode of development, in ideological
terms, involves the idea that the merits of socialism and the legacy of socialist revolution are to be more effectively defended and possibly increased
only through economic growth, which requires, in practical reality, engagement with global markets. *Understanding Sŏn'gun Politics* and other North
Korean analytical literature published in recent years about world affairs
and North Korea's place in the world, interestingly, are virtually silent about
the theory of market socialism and the related ideas of protecting socialist
revolution through engagement with the global economy. The absence of
any mention of this important politico-economic trend is understandable,
considering that the theory of market socialism goes head-on against the
military-centred theory of socialist revolution in the post-Cold War world
postulated by North Korea's *sŏn'gun* politics. Nevertheless, this absence
makes it an unsettling experience to read North Korea's recent political
literature, particularly because market socialism concerns developments in
countries that are historically and culturally closest to North Korea, and remains a haunting element in its discourse about North Korea's vanguard
role in global socialist revolution.

More can be said about the incongruity between North Korea's military-
first socialism and the economy-first socialism pursued by other Asian socialist polities, especially China, North Korea's close neighbour and traditional ally. Important in this matter is the text's allusion to North Korea's es-
teemed status in the Third World movement for political independence and
self-determination. After the Korean War (1950-1953), North Korea's polit-

ical development was complicated not only by the aggravated hostile relations with South Korea and the United States but also, since the late 1950s, by increasing hostility between the country's two principal allies and supporters of the war effort, China and Soviet Russia. In this turbulent milieu, North Korea sought to confront both of these separate hostile international environments by actively joining the non-aligned movement among the newly independent Third World nations and with the ambition to become a leader in the movement. The regional disputes between China and the Soviet Union and their rivalry in the 1960s and 1970s for global leadership in the socialist revolution played a major part in the rise of North Korea's strongly nationalistic, postcolonial state ideology, called *juch'e* ideology. This ideology replaced the largely Soviet-style, Soviet-allied political orientation in the earlier state-building era. According to Don Oberdorfer, "Beyond its sanctification of Kim [Il Sung]'s decisions, *juch'e* was a declaration of political independence from his two communist sponsors. Although it was originally called 'a creative application of Marxism-Leninism', eventually all reference to Marxist connections was abandoned." Oberdorfer (p.20, 1999). Subsequently, in the 1970s and early 1980s, North Korea made a considerable effort to elaborate upon this ideology and to export it to other developing nations in the Third World as a guiding ideology for national liberation and self-determination whose global relevance would be superior to that of the Leninist or Maoist precedents. North Korea's vigorous engagement with the non-aligned movement and its related ambition to become a global revolutionary power in the postcolonial world, equal to China or Soviet Russia in esteem and authority if not in military or economic power, are vividly demonstrated in one of North Korea's cherished public memorials, the International Friendship Exhibition Hall, which displays gifts to Kim Il Sung from all corners of the world.

In this regard, we may understand the idea of North Korea being the only remaining bastion of socialism. This is alluded to in the theory of military-first politics, as having a deeper historical background than the implosion of the Soviet political order or the subsequent isolation of North Korea. Tatiana Gabroussenko notes the following in her careful reading of changes in early North Korean literature:

> [From the late 1950s], North Korean policy makers started to position the DPRK as a self-sufficient state, equipped with a potentially world-dominating ideology. There emerged an image of North Korea as the sole, independent center of all the truly progressive forces of the globe. This image was strongly reminiscent of the image of the

USSR in old Soviet propaganda, with Pyongyang replacing Moscow as the centre of the inhabited universe. (p.43, Gabroussenko 2010).

Gabroussenko explores this image based on a dramatic shift in North Korea's literary production from a strong dependence on Soviet literature during the earlier postcolonial era of the 1940s and 1950s to a bold nationalist assertion for autonomy and authenticity beginning in the 1960s. She terms this development North Korea's "national Stalinism," highlighting the fact that the development was strongly prompted by the North Korean leadership's discontent with the dynamics of de-Stalinisation in the Soviet Union in the latter part of the 1950s. The challenge to Stalin's legacy in Soviet Russia posed a grave threat to the authority and esteem of North Korea's exemplary personality Kim Il Sung. Thus, we may say that the contemporary idea of North Korea as the single leader of the progressive world is rooted in North Korea's long-held ambition to be distinct from and treated as equal to its powerful neighbours, Soviet Russia and China. Just as *juch'e* ideology was supposed to mark North Korea as an original, independent player in global socialist and postcolonial politics during the earlier era, the theory of military-first politics is meant to set today's North Korea against the former Soviet Russia, which North Korea saw as a failed guardian of global revolutionary ideals. It also works to distinguish North Korea from China, which has concentrated, at least until recent times, on generating national economic growth rather than on sustaining the socialist and postcolonial revolutionary ideals, although this distinction is expressed in a much more implicit way than is the case with the former Soviet Russia.

The difference between North Korea's military-focused revolutionary socialist politics and China's economy-focused market-socialist politics points to other important issues. In his seminal work on early political and social revolution in North Korea, Charles Armstrong makes the important observation that North Korea's socialist revolution, compared to other socialist societies, was unique in its strong emphasis on correct thoughts and ideological dispositions as the driving force of revolution (rather than appropriate material conditions and an apt economic basis, as is the case in orthodox Marxism). Armstrong associates the idealistic, "neo-traditionalist" orientation of the North Korean socialist revolution partly with Korea's long, strong neo-Confucian tradition, which is considered to place supreme value on programmatic ritual propriety and nominal moral and ethical principles (pp.71-74 Armstrong 2004). North Korean literature on the origin of military-first politics indeed tends to stress the moral imperative to continue the socialist revolution and preserve the revolutionary heritage, privileging this imperative over questions of economic welfare and growth. In this re-spect, it is interesting to note that the term "preceding theories" appears frequently in North Korea's recent political and philosophical discourses (pp.21-22 Chun 2004). This term is intended to explain the originality and inventiveness of the country's guiding principles in relation to the tradi-tional Marxist and Leninist philosophical heritages. Kim Jong Il said the following in his speech delivered at the Workers' Party Central Committee meeting on 10 October 1990:

> The historical materialism of Marxism divided society into infras-tructure and superstructure, assigning the determining meanings to the former. This historical-materialist view is unable to establish a proper theory about the *juch'e* [meaning "subjectivity" in this con-text] of revolution... The Workers' Party is a political organisation assembled by people, who are also the object of this organisation's work. Therefore, the Party's organisational principles must be based on scientific understandings about the nature of human beings. All human activities are determined by thoughts and ideologies; there-fore, we must follow the principle of idealism in the construction of the Party.[5]

The self-conscious disassociation of North Korea's state ideology from the established socialist intellectual and revolutionary ideology was already in motion, according to Oberdorfer, in the earlier advancement of *juch'e* ideology in the 1960s (p.20 Oberdorfer 1999). *Juch'e* ideology advocates, as mentioned in the above quote, the centrality of human subjectivity in revolution – the idea that the collective human will is the main motor of historical progression – and accordingly makes claims against the existing theories of socialism and socialist revolution that privilege material forces and economic relations over ideational human moral and spiritual qualities. The idea of *juch'e* purports the history-making power of human collective subjectivity; it also centres this power on an extraordinary, exceptional and exemplary human subjectivity. Kim Jong Il, referring tacitly to the critique of personality cults in Marx's writings, says the following:

> The preceding theories approached the question of supreme leader-ship [*suryŏng*] merely as matters concerning the role of a superior

---

[5]Kim Jong Il, *"Jucheŭi dangkŏnsŏlironŭn rodonggyegŭpŭi dangkŏnsŏlesŏ t'ŭlŏjuigonaagaya hal jidojŏk jich'imida"* (The *juch'e* theory of party construction is the guiding principle for the construc-tion of the party of the working class) (speech delivered at the Plenum of the Central Committee of the Workers' Party, 10 October 1990).

individual [in revolution]. This problem also relates to the way in which preceding theories understood the role of a supreme leader mainly as questions of leadership. Questions about the supreme leader's status and role are not merely about his leadership. Instead, these are about the centre of socio-political organism, about this organism's highest cerebral organ. (p.20 Oberdorfer 1999).

This theory of political organism merits further attention. For the purpose of this essay, it suffices to mention that in the contemporary drive to advance the premises of military-first politics to a globally meaningful vanguard ideology, the claimed limitation and error of "preceding theories" refer specifically to how these theories conceptualize the relationship between military and economic forces in socialist revolution and state politics. *Understanding Sŏn'gun Politics* explains:

> The preceding theories prioritised relations of production and economic development, thereby advancing the idea that socialism and the socialist army can be built and strengthened in a country only when this country has a proper economic basis. By contrast, the military-first politics provides a new formula for revolution, based on the principles of *juch'e* ideology, which prioritises the empowerment of armed forces (pp.22-23 Chun 2004).

Based on this assertion that military-first politics and ideology constitute a decisive conceptual break with existing theories of socialist revolution, the text proceeds to an audacious assault on the premises of historical materialism and the primacy of material conditions in relation to ideological representation, turning the order practically upside down:

> The preceding theories advanced the working class as the principal army [of socialist revolution] based on socio-economic relations and conditions... These relations and conditions do not determine the principal forces of revolutionary movement; the determining force of socialist revolution is, rather than economic conditions, the people's ideological-mental qualities and their collective will and force. Military-first politics defines the revolutionary army as the core, primary power for the advancement of socialism and people's self-determination. (p.27 Chun 2004).

The theory of military-first politics privileges the power of ideology and the power of the army over the forces of production and, therefore, turns the principles of historical materialism and Marxism on their head.

Today

The North Korea of today is not the same place as the North Korea of yesterday. The formal state system remains largely unchanged, especially the centrality of the paramount leadership, although there has been some change since the current third-generation leadership took power in 2011, especially in the pursuit of military-first politics and related claims for a nuclear-armed statehood. Changes are found rather in the relationship between the state and the society – most notably in the general collapse of the state-distribution system of food and other basic subsistence goods. North Korean society today can no longer depend on the state for its physical survival, and the state has long stopped being the paternalistic guarantor of its citizens' economic well-being, despite the fact that in the rhetorical sphere, it has never been anything short of it throughout its seventy years of existence.

Important changes are also notable in the international environment. The so-called strategic competition between the United States and China has been made more remarkable and explicit in East Asia and beyond, and this has recently taken on an increasingly military character in addition to the largely economic and technological aspects in previous years. This is bound to bring changes to North Korea-China relations, as the texture of North Korea's military-first politics, defined previously in distinction to China's economy-first socialism, changes in relation to the escalating contest of power between China and the United States. Since Russia's invasion of Ukraine, Pyongyang has been making active diplomatic gestures towards Moscow, with strong rhetorical support for the latter. Observers speculate that in these precarious situations, Pyongyang may feel less isolated than before and find a niche and a new role in the emerging and radicalising great power politics.

Will North Korea still be able to claim the bright and shining status of the sole and solitary vanguard revolutionary power in the new environment as it did during the past three post-Cold War decades? Evidence suggests that Pyongyang is toning down this rhetoric of solitude somewhat while highlighting instead the virtue of friendship with China and Russia on the front of international solidarity against American power. If this modification finds a firmer ground (that is, if the great power politics continue and become more intense), we can safely anticipate a new manifesto on military-first politics sometime soon, in which a renewed international solidarity will be emphasised – that is, rather than the destiny of solitary resistance, as was the

Heonik Kwon

case in *Understanding Sŏn'gun Politics* on the place of North Korea in the post-Cold War world. The problem is, however, that none of this can make North Korea's solitude a thing of the past. To be a true and truly meaningful charismatic leader, the Leader must be not merely the leader of the nation but also of the world. This was the case with the country's founding leader (as exemplified in the exhibition of gifts to him), and it is equally the case with its hereditary successor and new keeper of the founding leader's legacy.

I look forward to learning how North Korea's dedicated political theories will resolve this contradiction and formulate a new argument –the contradiction between being a weak and marginal actor in the great power politics, on the one hand, and, on the other, the imperative of its own leader, not those of great powers, being the most bright-shining, singularly majestic player in it.

References

Armstrong, C. K. (2004), *The North Korean Revolution, 1945–1950*, Cornell University Press, Ithaca, U.S.A.

Chun, S.-P. (2004), *Sŏn'gun jŏngchie daehan lihae (Understanding sŏn'gun politics)*, Pyongyang Press, Pyongyang, North Korea.

Cummings, B. (2004), *North Korea: Another Country*, The New Press, New York, U.S.A.

Gabroussenko, T. (2010), *Soldiers on the Cultural Front: Developments in the Early History of North Korean Literature and Literary Policy*, University of Hawii Press, Honolulu, U.S.A.

Halliday, J. (1981), 'The North Korea enigma', *New Left Review* **127**, 18–52.

Kwon, H. (2010), 'North Korea's politics of longing', *Critical Asian Studies* **42**, 3–14.

Leffler, M. P. (2007), *For the Soul of Mankind: The United States, the Soviet Union, and the Cold War*, Hill and Wang, New York, U.S.A.

Oberdorfer, D. (1999), *The Two Koreas: A Contemporary History*, Warner Books.

Weber, M. (1947), *The Theory of Social and Economic Organization*, Free Press, ed. T. Parsons, New York, U.S.A.

8 Isolation in international relations

AMRITA NARLIKAR
German Institute for Global and Area Studies

Abstract: Since the end of the Second World War, diverse aspects of International Relations – including foreign policy, global governance, negotiation studies, and political economy – have been guided by an understanding that if markets were kept open, and states and their peoples interconnected, both prosperity and peace would stand a far better chance. In contrast, isolation, or its translation into a national strategy, isolationism, is often treated as a profanity in both the study as well as the practice of international relations. I offer a different perspective, that universal interconnectedness can no longer be treated as a rule-of-thumb for securing motherhood and apple-pie. Indeed, sometimes, time-bound or party-specific forms of isolation may be just the cure for certain types of political and economic maladies as well as to achieve new and updated goals. Illustrations are provided of cases where a move away from interconnectedness and towards some isolationism may be advisable, perhaps even necessary. The argument generates some interesting implications for research and policy.

> "...how sternly it's instilled,
> *All solitude is selfish.* No one now
> Believes the hermit with his gown and dish
> Talking to God (who's gone too); the big wish
> Is to have people nice to you, which means
> Doing it back somehow.
> *Virtue is social.*"
> - Philip Larkin, Vers de Société[1]

Since the end of the Second World War, diverse aspects of international relations (IR) – including foreign policy, global governance, negotiation

---

[1] From *Collected Poems* (Larkin 2003)

141

studies, and political economy – have been guided by an understanding that if markets were kept open, and states and their peoples interconnected, both prosperity and peace would stand a far better chance. World leaders and technocrats, often cheered on by like-minded scholars, have had a penchant to promote globalisation – freer flows of goods, services, capital, people, and ideas – with increasing interdependence being seen as a conduit to affluence, mutual understanding, and harmony among all nations. In contrast, isolation, or its translation into a national strategy, *isolationism*, is often treated as "a political slur" (Johnstone 2011) in both the study as well as the practice of IR.

I start off by highlighting the logic of the dominant aversion to isolationism in international relations. This approach has been a useful one; decades of consequent globalisation generated positive outcomes at multiple levels. The section that follows, provides instances of an unravelling of the happy bargain in recent years. I then argue that universal interconnectedness can no longer be treated as a rule-of-thumb for securing utopia; indeed, sometimes, time-bound or party-specific forms of isolation may be just the cure for certain types of political and economic maladies as well as to achieve new and updated goals. I provide illustrations of cases where a move away from interconnectedness and towards some isolationism may be advisable, perhaps even necessary. The concluding section draws out the research and policy implications of my argument.

"When hope and history came to rhyme"[2]

Economists with impeccable credentials, going back to Adam Smith, had argued for trade openness across nations. Moreover, the gains from cross-border exchange were not expected to be limited to the economic realm. Writing in 1795, German philosopher Immanuel Kant identified three necessary conditions to achieve a "perpetual peace" among nations (Kant 1795). The first was a liberal, representative, democratic form of government within countries, which would ensure that governments acted in the interests of the majority. The second involved a recognition of human rights, which transcended national differences. Together, the first two would contribute to trust and non-aggression among fellow republics. And then came the third condition, the most interesting from the perspective of this Darwin Lectures' theme, an openness across this community of republics. Michael

---

[2]Quote from (Heaney & Sophocles 2018)

142

Doyle summarises this last principle as follows: "trade, tourism and other forms of transnational contact grow which lead to prosperity, reinforcing mutual understanding with many opportunities for profitable exchange, and producing contacts that offset in their multiplicity the occasional sources of conflict" (Doyle 2000). Exchange among diverse nations and peoples, the opposite of isolationism, thus formed a key part of the Kantian promise of peace.

Real world events in the 20th century seemed to reinforce the Kantian message. Faced with the Great Depression in 1929, countries turned inwards and adopted beggar-thy-neighbour policies, thereby deepening the severity of the economic crisis as well as its political consequences. Policymakers took these tough lessons seriously as they devised plans for a new, post-war, world order. Henry Morgenthau, US Secretary of Treasury, addressed the inaugural plenary session (1944) of the Bretton Woods Conference with the following words:

> "All of us have seen the great economic tragedy of our time. We saw the world-wide depression of the 1930s. We saw currency disorders develop and spread from land to land, destroying the basis for international trade and international investment and even international faith. In their wake, we saw unemployment and wretchedness, idle tools, wasted wealth. We saw their victims fall prey, in places, to demagogues and dictators. We saw bewilderment and bitterness become the breeders of fascism and, finally, of war.

> In many countries controls and restrictions were set up without regard to their effect on other countries. Some countries, in a desperate attempt to grasp a share of the shrinking volume of world trade, aggravated the disorder by resorting to competitive depreciation of currency. Much of our economic ingenuity was expended in the fashioning of devices to hamper and limit the free movement of goods. These devices became economic weapons with which the earliest phase of our present war was fought by the Fascist dictators. There was an ironic inevitability in this process. Economic aggression can have no other offspring than war. It is as dangerous as it is futile.

> "We know now that economic conflict must develop when nations endeavour separately to deal with economic ills which are international in scope. To deal with the problems of international exchange and of international investment is beyond the capacity of any one

country, or of any two or three countries. These are multilateral problems, to be solved only by multilateral cooperation." (Morgenthau 1944).

The post-war agenda for global governance was thus clear: both the self-interest of states and the public good would be well-served by deepening economic interconnectedness and interdependence. A range of international relations theories provided further inspiration to shape and reinforce this approach: neo-liberal institutionalism, functionalism, complex interdependence, constructivism all argued (using different mechanisms) that institutionalised and sustained patterns of interdependence and cooperation would result in increasing regime compliance and even "socialisation". At different points of time and for different reasons (ranging from changes in technology, balances of power, domestic politics, as well as in the prevailing epistemic consensus at the time) emphasis would be placed on different types of trade-offs within the globalisation bargain[3]. But the general narrative – across the political spectrum – was one that rooted for closer economic ties across countries, with a potential for positive spillovers for further cooperation and growing integration across other areas as well.

This logic underpinned the multilateral institutions of the post-war era. Recall, for instance, that the European Union, with its guarantee of the four freedoms (movement of goods, people, services and capital), the European Court of Justice, common currency of the euro, competition policy, and further aspects of sovereignty pooled among 27 members, had its humble origins in the European Coal and Steel Community among the 6 countries of Belgium, France, Germany, Italy, Luxembourg and Netherlands (Vertrag der Begrundung, Vertrag über die Gründung der Europäischen Gemeinschaft für Kohle und Stahl 1952). The foundation stone of the multilateral trading system, the General Agreement on Tariffs and Trade (GATT), was similarly a limited agreement (as the name itself suggests)[4]. Over the decades, the regime came to include behind-the-border measures (such as Technical Barriers to Trade and Sanitary and Phytosanitary Barriers to Trade), and expanded its coverage to services, Trade-Related Intellectual Property Rights, and Trade Related Investment Measures. In 1995, the World Trade Organization (WTO) was created – an "organisation" with

---

[3] e.g. Martin Daunton, 'The Inconsistent Quartet: Free Trade versus Competing Goals' in (Daunton et al. 2012), and (Obstfeld et al. 2005)

[4] On the history of the multilateral trading regime including the failed experiment to create a more expansive International Trade Organization in the 1940s, see (Narlikar 2020)

Amrita Narlikar

"members" (in contrast to the shallower integration envisaged under the umbrella of the GATT and its "contracting parties", and also a much stronger Dispute Settlement Mechanism. From the 23 founding contracting parties to the GATT, to the 164-strong World Trade Organization, the multilateral trading regime has come a long way; this travel, moreover, has been in the direction envisaged by the anti-isolationist proclivities of the Bretton Woods founding fathers.

The integration of global markets produced many gains across the board. Trade openness, for instance, contributed significantly to poverty reduction. A 2015 report, produced jointly by the World Bank and the World Trade Organization, noted: "The number of people living in extreme poverty around the world has fallen by around one billion since 1990. Without the growing participation of developing countries in international trade, and sustained efforts to lower barriers to the integration of markets, it is hard to see how this reduction could have been achieved." (*The Role of Trade in Ending Poverty* 2015) The emergence of the BRICs – Brazil, Russia, India, and China – as growth markets formed a part of the transformation that seemed underway.

The gains, moreover, went beyond the economic realm; countries from the Global South also came to acquire more voice in international organisations. In a world of increasing global interconnectedness that was managed by multilateral rules, small and poor countries were able to form coalitions and use the global stage (provided by international organisations) to push forwards their own concerns. The launch of a new round of trade negotiations that was dedicated to development concerns in 2001, under the umbrella of the WTO, was a shining example of the power that poor countries had come to acquire (Narlikar 2020).

At a normative level too, a freer flow of images and ideas was generating empathy and support across different parts of the world. Our digital connectivity has created unprecedented opportunities for worthy causes to go viral. The instruments for action have also become more accessible: we are not only more aware of the sufferings of other humans and animals through the images that stream across diverse media and social media channels, but also have the possibility to join forces with other activists (digitally and in person) and contribute with donations with a swift press of a button to help those who are thousands of miles away. The fact that schoolchildren worldwide can come together for climate action via Fridays for Future[5] protests

---

[5] A global movement involving children and young adults, whose primary agenda is to mobilise action to address the climate emergency, www.fridaysforfuture.org

is, in good measure, a product of this interconnectedness. None of these developments would have been possible in a world of isolation or fragmentation. The hopes of those who had so carefully developed the post-war global order of interconnectedness were rhyming beautifully with history as it unfolded in the second half of the 20th century. Globalisation seemed to have poetry in it.

"Things fall apart; the centre cannot hold"[6]

A closer look at even the "poetic" decades of globalisation (in the post-World War II era), and it was obvious that globalisation had had its discontents. Developing countries, for instance, had tenaciously complained of exclusion and marginalisation from multilateral organisations; they had sought reform from within the established institutions (including via coalitions in the GATT and the G77 in the United Nations) and also tried to build alternative organisations (such as the United Nations Conference on Trade and Development in 1965), albeit only with limited success (Stiglitz 2003, Chang 2002, Narlikar 2004). The system was also not universal; for instance, most countries of the former socialist bloc (bar the then Yugoslavia and Cuba) were not contracting parties to the market-oriented GATT. But with the end of the Cold War, the system became more inclusive. Countries from the former Eastern bloc began to join various international organisations, ranging from the European Union to the WTO. Developing countries also began to acquire greater voice, as highlighted in the previous section. But even as a universal, liberal utopia of interconnectedness was taking shape, fissures began to show and grow.

For all the promise of further globalisation, the Doha Development Agenda – the first trade round in the WTO to be explicitly dedicated to the concerns of the Global South – started to run into recurrent deadlocks from 2003 onwards and eventually whimpered to an inglorious end[7]. In 2016, apparently convinced by the argument of "take back control" by the Brexit campaign, the British electorate voted to leave the European Union. In 2017, Donald J. Trump assumed power as the President of the United States. The president threw a major spanner in the works of global trade by declaring trade wars as being "good and easy to win" and then unilaterally impos-

---

[6]Quote from *The Second Coming*, Yeats (collected works, 2000)

[7]A recent and rather limited deal achieved at the WTO in June 2022 is a far cry from the ambitions of the Doha Development Agenda (Narlikar 2022)

ing tariffs against a variety of partner countries. He also unreservedly lambasted the multilateral rules that underpinned global order (e.g. "the WTO is the single-worst trade deal ever made" (*'The single worst trade deal ever made': Trump threatens to remove US from WTO* 2018)). Previous US presidents had also expressed their dissatisfaction with the workings of the world trade order, and held up the appointment/re-appointment of members of the Appellate Body to the WTO's Dispute Settlement Mechanism. Under the Trump administration though, this mechanism was actually paralysed due to the refusal of the US to appoint/renew the Appellate Body's members; this state of stasis continues at the time of writing. If anything, and in spite of President Joe Biden's cheery announcement of "America is back!", the US has continued with many of Trump's strategies on trade. The same applies to the prioritisation of domestic requirements over international commitments[8]. Such disillusionment with the existing patterns of globalisation and governance, prompting a turn inwards – and that too coming from the world's largest economy, which had had a major role in building the global order – was serious. The rise of populism across different parts of the world, including Europe, highlighted the growing dissatisfaction of voters with the existing bargains on globalisation (for example Rodrik (2018)). Demonstrations and riots that came to accompany much global summitry were another indication of this[9]. And the challenge did not stop there. Authoritarian advance has been on the rise, whether in the shape of the Russian invasion of Ukraine or Chinese adventurism in its neighbourhood, flying in the face of the optimism of policy-makers[10].

Supporters of the existing economic order are at pains to remind us that despite all the doom-saying, globalisation has continued: "So far, the fabric of globalisation has proved so densely woven it has resisted attempts at unravelling" (Beattie 2022, Lamy & Köhler-Suzuki 2022). But the political and economic realms are out of sync, often displaying diverging tendencies towards isolationism and globalisation respectively; meanwhile, the world has become much more fractious, violent, and dangerous. All these developments on the ground should give us pause for thought about the current

---

[8]It is worth noting that neither Trump nor Biden is unique among recent US presidents to adopt these positions; President Obama had also indicated a tendency on the part of the US to disengage from its international role, see Goldberg (2016)

[9]The G20 summit in Hamburg in 2017 was a case in point (Narlikar 2017).

[10]An example of this optimism could be found in the idea of *Wandel durch Handel* in Germany, literally change through trade. This translated more broadly into the expectation that by economically engaging China, Russia, and others, we would see greater norm convergence of authoritarian states towards liberalism, democracy, and peace

model of globalisation, and the assumptions of isolation versus interconnectedness that underpin it.

"When the facts change, I change my mind, what do you do, sir?"[11]

The challenge to the current model of globalisation has several origins.
For decades, some critics point out, the gains of globalisation have been
unequally distributed (especially within countries). Further, technocratic
hubris has led defenders of globalisation to assume that the gains of interconnectedness are obvious to all parties, and become complacent in developing inclusive narratives about globalisation that can continue to win the
hearts and minds of their populations. Nationalistic narratives that promise
to defend the interests of one's own electorates (such as "Make America
Great Again" by Trump) have had better uptake than those that commit to
providing a global public good (such as "Make Our Planet Great Again"
by the French President, Emmanuel Macron in 2018, which prompted the
Yellow Vests' protests) (Narlikar 2020). But additionally, and perhaps even
more worryingly, a fundamental structural change has taken place that un-
dermines, sometimes even overturns, the assumed correlation between economic interconnectedness and peace: the possibility that some states can
"weaponise" interdependence to the considerable disadvantage of others.

In their pioneering work on "Weaponised Interdependence", Henry Farrell and Abraham Newman point to the transformative impact of changing
production patterns on global power balances (Farrell & Newman 2019).
Trade and production today are organised around global value chains; these
networks are not flat (contrary to what previous studies had assumed e.g.,
Friedman 2005, Slaughter 2009, 2018) but highly asymmetric. Only a few
states have political authority over network hubs, and the necessary domestic institutions that can enable them to exploit their strategically privileged positions. Through two mechanisms that Farrell and Newman identify, panopticon and chokepoint effects hub states are able to "extract informational advantages vis-à-vis adversaries" and also "cut adversaries off
from network flows". Through these effects, a few powerful states can "discover and exploit vulnerabilities, compel policy change, and deter unwanted

---

[11] This trenchant remark has been attributed variously, and perhaps apocryphally, to luminaries including Winston Churchill and John Maynard Keynes

Amrita Narlikar

actions."[12]

Besides the structural logic of global production patterns, different political systems do allow some states greater institutional capacity to exercise control over network hubs, and make more effective use of the panopticon and chokepoint effects. This matters more than ever today. Most countries of the former eastern bloc were, for instance, not contracting parties to the GATT; the system now includes states such as China, with different state-market relationships and bargains (Wright 2013). These differences in political systems, and varying levels of state control over private authority and network hubs that they enable, do not make for a level playing field. High network externalities further create significant barriers to entry for new players; monopolies emerge around some key supply chains that tend to reinforce and exacerbate existing power asymmetries (Farrell et al. 2021).

Under conditions of weaponised interdependence, the logic of the post-war system gets turned on its head because interconnectedness, originally seen as the instrument of peace and convergence, now offers a handful of already powerful states the possibility of unprecedented control over others. The post-war system of global governance was not built for a world where interdependence itself could be weaponised. In this brave new world, an ideological commitment to universal interconnectedness will not serve us well. And we are already seeing the effects of weaponised interdependence in key aspects of our lives.

FOOD SECURITY

Food security is a case in point. The old model of globalisation, which had equated food security with international trade supposedly to be based on well-functioning supply chains, has now come to bite. There are multiple illustrations of, and reasons for, the problem of hunger and food insecurity. Let us consider a recent illustration, stemming from events close to home: i.e., Russia's war against Ukraine and its global repercussions for food supply and prices. First, with the Russian invasion of Ukraine (both

---

[12]Note that in the past too, some states were able to use and abuse the power of their large markets as well as exploit bilateral dependence (e.g. Keohane & Nye Jr 2011). As Farrell and Newman show, however, weaponised interdependence operates at a different level: "Global economic networks have distinct consequences that go far beyond states' unilateral decisions either to allow or deny market access, or to impose bilateral pressure. They allow some states to weaponise interdependence on the level of the network itself".

"grain exporting powerhouses that accounted for 24% of global wheat exports by trade value, 57% of sunflower seed oil exports and 14% of corn from 2016 to 2020, according to data from UN Comtrade" (Bankova et al. 2022)), already high food prices were expected to increase further due to the obvious disruption that the war would produce – and have indeed shown this pattern. Second, the sanctions against Russia themselves seem to have had unintended effects on the country's food exports, prompting a reconsideration of a part of the sanctions package by the EU (Guarascio 2022). Russia's blockade of ports has been an effective weaponisation of trade routes, which hurts Ukraine and also holds a major chunk of the world's food supply hostage (Snyder 2022). A much-touted deal was brokered by Turkey and the United Nations on 23 July 2022, through which both Russia and Ukraine agreed to reopen Black Sea routes for food exports; before the deal was even a day old, an attack on Odesa (attributed to Russia but denied by the Russian government) put the deal in jeopardy. The multilateral trading system has not been able to help as different countries put up formal and informal restrictions on food exports[13]. Least developed countries find themselves in a particularly precarious situation as a result of these shortages and higher food prices, but middle-income and rich countries are also facing growing discontent among their own populations. A similar pattern is also evident with regard to energy supply chains. Germany, for instance, finds itself torn: its goal of defending Ukraine's sovereignty via sanctions against Russia is proving difficult to reach, given its long-standing over-reliance on Russia to meet its own oil and gas needs. Diversification takes time and is costly.

PANDEMIC POLITICS AND COOPERATION FAILURES

A second important example derives from our recent experience with the COVID-19 pandemic. Global collaborations among scientists on understanding the disease itself, and developing vaccines plus other treatments, have rightly been lauded as convincing illustrations of the virtues of interconnectedness. But there are also some dark sides to this story, not least the tragic death of the "Wuhan whistleblower" Dr Li Wenliang (*Li Wenliang:*

---

[13]Reuters Staff e.g., 2022, Menon e.g., 2022. After much haggling in Geneva in June 2022, WTO members agreed not to "impose export prohibitions or restrictions on foodstuffs purchased for noncommercial humanitarian purposes by the World Food Programme" (Ministerial Decision, WT/MIN(22)/29, WT/L/1140, 17 June 2022); this measure, however, is extremely limited and will not suffice to address the problem of food insecurity Calvo e.g., 2022

Amrita Narlikar

'Wuhan whistleblower' remembered one year on 2021). Recall the failures of international cooperation in the early stages of the pandemic, e.g.:

> "It is surprising, then, that when a pneumonia-like virus was detected in Wuhan in late-December 2019, the WHO, armed with data and years of subsequent research about the SARS outbreak, reacted as sluggishly as it did. Dr Tedros Adhanom Ghebreyesus, better known as Dr Tedros, the Director General of WHO, applauded China's "commitment to transparency" in the early days of the epidemic in January, despite mounting evidence to the contrary. The WHO then denied evidence of human-to-human transmission of the novel coronavirus, barely a day after the first case was announced outside China. This is despite the fact that Taiwan, whose exclusion from the WHO deserves an article in itself, had warned the body of this as early as December.
>
> While Beijing informed the WHO on December 31, there are expert estimates that the virus had spread to humans as far back as October. Even after being told, the WHO showed no urgency to send an investigative team, careful not to displease the Chinese government. A joint WHO-Chinese team went to Wuhan only in mid-February and wrote a report with decidedly Chinese characteristics."[14]

This sorry state of affairs continued when it came to life-saving medical supplies and equipment during the early months of the pandemic. Masks were in short supply as were tests, personal protective equipment, and basic drugs. Some economists have argued that the cause of these problems lay in inadequate stockpiles, rather than disruption to trade flows (Evenett & Baldwin 2020). But it is important to bear in mind that the old model of globalisation has been premised on the (seemingly) dependable efficiencies of global trade, under which countries are actively discouraged from stockpiling essential or strategic products. When faced with shortages, not only did countries put up export restrictions on key medical supplies, but they also used medical surpluses as bargaining chips[15]. Despite there being capacity for production available in the developing world, and a desperate

---

[14] Saran 2020, For further accounts of the sins of omission and commission on the part of the WHO as well as China, see Anne Applebaum, 'When the World Stumbled: COVID-19 and the Failure of the International System', and Thomas Wright, 'COVID19's Impact on Great Power Competition', in Brands 2020

[15] Amrita Narlikar, 'Must the Weak Suffer what they Must: The Global South in a World of Weaponized Interdependence' in Farrell et al. 2021

demand for access to vaccines, big pharmaceutical companies and developed countries have been reluctant to offer a relaxation of the Trade-Related Aspects of Intellectual Property Rights regime.[16] Interdependence can have life and death consequences, if the existing system of rules fails to ensure equal access to life-saving medicines.

OPEN BORDERS DURING A PANDEMIC

While the previous two examples of food security and cooperation failures during the pandemic suggest a re-alignment of production chains (e.g. with like-minded partners, with whom a reliable supply is more certain), we also have recent examples of the importance and value of isolation. Writing about China, for instance, Thomas Wright argues, "In February 2020, it put considerable pressure on countries not to restrict travel with China, even as it prohibited domestic travel to and from Wuhan domestically. It asked donor countries to keep a low profile to save face. Later, it would freely impose its own travel restrictions on other countries. . . (Brands 2020). China was not the only country fighting against border closures for its citizens. The Europeans took an almost dogmatic stance in favour of open borders, with much finger-pointing when countries unilaterally imposed border controls and exercising considerable peer pressure to bring about early reopenings. Tracing the early trajectory of the pandemic in Europe, Cecilia Sottilotta writes, "Although synchronous (if not always timely) lockdowns eventually allowed the highly integrated continent to 'flatten' the contagion curve during the first wave (February-May 2020), more open borders (and commitment among EU authorities to keep them open) exposed countries to further outbreaks" (Sottilotta 2022).

Supporters of the old model of globalisation tend to argue that these failures illustrate the necessity of *more* interconnectedness, not less. The argument would run: we need stricter trade rules such that Russia would not be able to place costly curbs on food exports; neither China nor any other country should be able to cover up the outbreak of a pandemic, for which the WHO needs to be given more powers; rules of economic exchange should infuse more reliability into global supply chains; get all these measures right and border controls might not be needed at all. But such counsel will be very difficult to implement short of the emergence of a utopian world

---

[16]The Geneva Package, reached in June 2022, is a case of too-little-too-late, e.g. (Okonjo-Iweala 2022), also (Statement 2022)

government; indeed, the past seventy years of global governance have tried to do this with varying degrees of success. A different, and more useful lesson from these recent experiences is that we need to take off our rose-tinted glasses, and build a new form of globalisation that is more restrained, mindful, and allows different levels of integration with different types of partners[17]. What could this alternative model look like?

Madmen in authority, academic scribblers, and isolationism for the study and practice of IR[18]

The old model of globalisation may have served us well for several decades, but the world has changed. An ideological adherence to universal inter-connectedness can be counter-productive, as the previous section has illustrated. The system needs a reboot, as do our academic models that underpin it. Below, I offer four suggestions on the next steps for researchers and practitioners.

First, as scholars, we are wont to complain that policy failures are a product of practitioners not heeding our sage advice; rarely, however, do we see the necessary reflection by experts on when and how they might have also gotten the situation quite wrong[19]. In this vein, international relations researchers would do well to question the relevance of an assumed dichotomy between interdependence/internationalism versus isolationism, which many have bought and also advised policy-makers on. Polarisation on the ground, in recent years, has admittedly reinforced this dichotomy: faced with the hostile and somewhat isolationist impulse (e.g. on the part of Trump or some Brexit campaigners) towards global governance institutions, defenders of multilateralism often end up in a knee-jerk reaction to protect

---

[17]E.g.: "We may need cooperation with China to tackle pandemic disease properly, but China's national and global response to Covid-19 should remind us that we should be realistic about how much cooperation we can get from China's Communist Party regime. It will be limited, imperfect, and hard to trust", for example Brands 2020; also Narlikar 2021 and Narlikar & Müller 2021

[18]The title for this section is of course drawn from an oft-cited quote by John Maynard Keynes: "Practical men who believe themselves to be quite exempt from any intellectual influence, are usually the slaves of some defunct economist. Madmen in authority, who hear voices in the air, are distilling their frenzy from some academic scribbler of a few years back" in *The General Theory OF Employment, Interest AND Money* 1936.

[19]This, in turn, contributes to a general distrust of experts, epitomised by Michael Gove's: "I think the people of this country have had enough of experts with organisations with acronyms saying what they know is best and getting it consistently wrong", Interview with Sky News, 22 June 2016. For a recent attempt to address problems at the interface of science and policy head on see 'The How Not to Guide for International Relations' Drezner & Narlikar 2022

an unreformed globalisation. In fact, a more useful way to think about interconnectedness and isolationism is not in terms of binaries, but along two ends of a conceptual spectrum. Depending on external conditions (including changing balances of power) and domestic preferences, it is at least in principle for states to choose at which point of the spectrum they wish to be located. This means, for instance, that the choice does not have to be between free trade or self-reliance; rather, the answer may lie in re-aligned supply chains that build on more domestic content production (within the bounds of feasibility and sustainability) and – even more importantly – closer trade relations with like-minded partners and allies, especially for strategically important sectors. In practice, however, these choices are not easy: our institutions of multilateral governance have restricted these choices, but so have the narratives of dichotomy by which we have allowed ourselves to become entrapped.

A second, lazy shortcut into the debate is an implicit assumption: to be in favour of multilateralism automatically implies that one must be against isolationism. The reason for this may derive partly from the fact that the multilateral institutions of the post-war era were proponents of globalisation, leading to important differences between the two – as concepts – to elide. Further, as most scholars of IR tend to have internationalist orientations, there is also a subconscious preference for multilateral cooperation across multiple theories and approaches of IR (with Realism being a prominent exception)[20]. These unspoken assumptions and preferences tend to reinforce the polarisation highlighted above, with key actors accepting roles that are caricatures of multilateralists (e.g. Angela Merkel, arguably, showed this tendency, and was duly applauded by an adoring western media) and isolationists (e.g., Donald J. Trump seemed to relish the role of being the party pooper across different multilateral parties). In fact, however, one can be fully committed to multilateral cooperation, and still build a system of rules that allows certain measures/types/durations of isolationism.

Third, while examples have been offered of when we need to accept isolationism, indeed embrace it in certain conditions, it is important to note that the reasoning does not advocate isolationism as a general policy. Integration, interdependence, globalisation, whichever version we take, have served us well. For economies to now turn inwards on a wholesale basis would do huge damage to the decades of poverty alleviation that have been achieved globally; we would all find ourselves less prosperous and also less

---

[20] I am grateful to Charlie Roger for a stimulating exchange on both these points.

Amrita Narlikar

secure. What has been suggested here, is finding a balance between across-the-board globalisation on the one hand (that can be weaponised), and complete isolation on the other (that could dramatically reduce development and prosperity for all parties). Trade-offs will be involved. Re-aligned production chains will bring greater certainty of supply as well as potentially enhance national security, but there will be short-term costs associated with decoupling/diversification, as well as longer-term costs of forgoing cheaper and less reliable suppliers. These trade-offs will have to be openly recognised, and negotiated with domestic populations and electorates. For instance, trying to wean Germany off Russian gas is proving to be costly, but the government is doing this nonetheless in anticipation of gains to security as well as support for Ukraine, which in turn can also be seen as a strategy towards enhancing the security of European democracy at large.

Finally, any experiments with isolationism will fail miserably if – paradoxically – implemented in isolation. Allies will be key to breaking out of the hazardous dependence that the current model of globalisation promotes. A new set of multilateral rules will be needed to prevent races to the bottom, and to achieve a new, sustainable bargain that allies and rivals can live with. In the often loud and stylised debate on decoupling, one is frequently presented with the bogeyman of a renewed Cold War. I would humbly suggest that our best chance of avoiding a new Cold War, or indeed a large-scale hot war, is to allow for time-bound and plurilateral escapes into isolationism, Fig. 8.1.

Figure 8.1 Rethinking isolation, illustration by Vladdo (Vladimir Flórez), inspired by Amrita Narlikar's work, who acknowledges Julia Kramer for her suggestions.

Amrita Narlikar

References

Bankova, D., Dutta, P. K. & Ovaska, M. (2022), 'The war in Ukraine is fuelling a global food crisis', https://graphics.reuters.com/UKRAINE-CRISIS/FOOD/zjvqkgomjvx/.

Beattie, A. (2022), 'No, the global economy is not breaking into geopolitical blocs, *Financial Times*', https://www.ft.com/content/d50ebceb-3179-4410-84ab-e8285cbcb644.

Brands, H., ed. (2020), *COVID-19 and World Order: The Future of Conflict, Competition, and Cooperation*, Johns Hopkins University Press, Baltimore, MD.

Calvo, F. (2022), 'How can the WTO contribute to global food security?', https://sdg.iisd.org/commentary/policy-briefs/how-can-the-wto-continue-delivering-good-outcomes-on-food-security/.

Chang, H.-J. (2002), *Kicking Away the Ladder: Development Strategy in Historical Perspective*, Anthem Press.

Daunton, M., Narlikar, A. & Stern, R. M. (2012), *The Oxford Handbook on The World Trade Organization*, Oxford University Press, Oxford, U.K.

Doyle, M. W. (2000), 'A more perfect union? the liberal peace and the challenge of globalization', *Review of International Studies* **26**(5), 081–094.

Drezner, D. & Narlikar, A. (2022), *International Affairs*, Vol. 98, Oxford University Press, chapter The How Not to Guide for International Relations, pp. 1499–1513.

Evenett, S. J. & Baldwin, R. (2020), *Revitalising Multilateralism Pragmatic Ideas for the New WTO Director-General*, The Centre for Economic Policy Research.

Farrell, H. & Newman, A. L. (2019), 'Weaponized interdependence: How global economic networks shape state coercion', *International Security* **44**, 42–79.

Farrell, H., Newman, A. L. & Drezner, D. W., eds (2021), *The Uses and Abuses of Weaponized Interdependence*, Brookings Institution Press.

Friedman, T. L. (2005), *The World Is Flat: A Brief History of the Twenty-first Century*, Farrar Straus & Giroux.

Goldberg, J. (2016), 'The Obama Doctrine', https://www.theatlantic.com/magazine/archive/2016/04/the-obama-doctrine/471525/.

Guarascio, F. (2022), 'EU to Soften Sanctions on Russian Banks to Allow Food Trade', *Reuters* .

Heaney, S. & Sophocles (2018), *The cure at Troy: A version of Sophocles' Philoctetes*, Faber & Faber.

Johnstone, A. (2011), 'Isolationism and Internationalism in American Foreign Relations', *Journal of Transatlantic Studies* **9**, 7–20.

Kant, I. (1795), *Perpetual Peace; a Philosophical Essay*, HardPress Publishing.

Keohane, R. O. & Nye Jr, J. S. (2011), *Power & Interdependence*, Pearson.

Keynes, J. M. (1936), *The General Theory of Employment, Interest and Money*, Palgrave Macmillan, London.

Lamy, P. & Köhler-Suzuki, N. (2022), 'Deglobalization is not inevitable, *Foreign Affairs*', https://www.foreignaffairs.com/articles/world/2022-06-09/deglobalization-not-inevitable.

Larkin, P. (2003), *Philip Larkin: Collected Poems, edited by A. Thwaite*, Faber & Faber, London.

Li Wenliang: 'Wuhan whistleblower' remembered one year on (2021), https://www.bbc.co.uk/news/world-asia-55963896.

Menon, S. (2022), 'India wheat export ban: Why it matters to the world', https://www.bbc.co.uk/news/world-asia-india-61590756.

Morgenthau, H. (1944), 'Inaugural Address at the Plenary Session of the Bretton Woods Conference', https://www.cvce.eu/content/publication/2003/12/12/34c4153e-6266-4e84-88d7-f655abf1395f/publishable_en.pdf.

Narlikar, A. (2004), *International Trade and Developing Countries: Bargaining Coalitions in GATT and WTO*, Routledge, London.

Narlikar, A. (2017), 'The Real Power of the G-20, *Foreign Affairs*', https://www.foreignaffairs.com/articles/world/2017-07-25/real-power-g-20.

Narlikar, A. (2020), *Poverty Narratives and Power Paradoxes in International Trade Negotiations and Beyond*, Cambridge University Press.

Narlikar, A. (2021), 'Emerging narratives and the future of multilateralism', *Raisina Files* pp. 1–11.

Narlikar, A. (2022), 'How not to negotiate: The case of trade multilateralism', *International Affairs* **98**, 1553–1573.

Narlikar, A. & Müller, N., eds (2021), *Making it Matter: Thoughts Experiments in Meaningful Multilateralism*, Körber-Stiftung & German Institute for Global and Area Studies (GIGA), Hamburg.

Obstfeld, M., Shambaugh, J. C. & Taylor, A. M. (2005), 'The Trilemma in History: Tradeoffs among Exchange Rates, Monetary Policies, and Capital Mobility', *Review of economics and statistics* **87**, 423–438.

Okonjo-Iweala, D. G. (2022), 'WTO members tout MC12 success; package gets mixed reviews from business, civil society', https://insidetrade.com/daily-news/okonjo-iweala-wto-members-tout-mc12-success-package-gets-mixed-reviews-business-civil.

Reuters Staff (2022), 'Bulgaria hampering grain exports amid ukraine war, producers say', https://www.reuters.com/article/ukraine-crisis-grain-bulgaria-idAFL5N2V60E6.

Rodrik, D. (2018), 'Populism and the economics of globalization', *Journal of International Business Policy* pp. 12–33.

Saran, S. (2020), '#covid19: Dr WHO gets Prescription Wrong', *Health Express, Observer Research Foundation* .

Slaughter, A.-M. (2009), *A New World Order*, Princeton University Press, Princeton, NJ.

Slaughter, A.-M. (2018), *The Chessboard and the Web: Strategies of Connection in a Networked World*, Yale University Press, New Haven, CT.

Snyder, T. (2022), 'Tweet', https://mobile.twitter.com/TimothyDSnyder/status/1543326437742198784.

Sottilotta, C. E. (2022), 'How not to manage crises in the European Union', *International Affairs* **98**, 1595–1614.

Statement, J. (2022), 'WTO kneels to 'big pharma' at ministerial conference', https://publicservices.international/resources/news/wto-kneels-to-big-pharma-at-ministerial-conference?id=13117&lang=en.

Stiglitz, J. (2003), *Globalization and Its Discontents*, Penguin, London.

The Role of Trade in Ending Poverty (2015), World Bank Group and World Trade Organization.

'The single worst trade deal ever made': Trump threatens to remove US from WTO (2018), https://www.smh.com.au/business/the-economy/the-single-worst-trade-deal-ever-made-trump-threatens-to-remove-us-from-wto-20180831-p500zl.html.

Vertrag der Begrundung, Vertrag über die Gründung der Europäischen Gemeinschaft für Kohle und Stahl (1952), https://eur-lex.europa.eu/legal-content/DE/TXT/PDF/?uri=CELEX:11951K/TXT&from=EN.

Wright, T. (2013), 'Sifting through interdependence', *The Washington Quarterly* **36**, 7–23.

Yeats, W. B. (2000), *The Collected Poems of W. B. Yeats*, Wordsworth Editions, Ware.

Notes on contributors

Christine van Ruymbeke (Université libre de Bruxelles, Ph.D) is Ali Reza and Mohamed Soudavar Professor of Persian Literature and Culture at the Faculty of Asian and Middle Eastern Studies at Cambridge. She is also Graduate Tutor and the College Praelector at Darwin College, Cambridge. (http://www.ames.cam.ac.uk/directory/vanRuymbeke).

She is a long-standing member of the councils of learned societies, including the Iran Heritage Foundation (London, UK), the Royal Asiatic Society (London, UK) and the Ancient India and Iran Trust (Cambridge, UK). She was elected Secretary to the Board of the Societas Iranologica Europaea in 2019 (https://www.societasiranologicaeu.org/board/). In 2009, her book *Science and Poetry in Medieval Persia. The Botany of Nizami's Khamsa*, received the World-prize of the Book of the Year of the Islamic Republic of Iran.

Christine is a literary critic working on classical (medieval and pre-modern) Persian literature, with a special focus on its medieval non-mystical narrative production. How can we engage today with these classical Persian texts? Can we regard them as meaningful and helpful despite the temporal and geographical distance with their authors? Christine approaches these medieval texts through aspects such as medieval authorial rewriting techniques and anxiety of influence; narrative structures and embedding techniques; intratextuality, poetics and cognitive effects, bringing together medieval and present-day literary and poetic theory. She explores how Persian poetry has irrigated other traditions well beyond the Persianate world, such as early-modern French and Victorian English literatures. This approach also encompasses the poetry's impact on visual culture, especially the calligraphy and illustrations in medieval and pre-modern Persian manuscripts.

Amy Nethery is a political scientist and Senior Lecturer at Deakin University. She researches the development and impact of asylum policies in Australia and Asia. She has a particular interest in immigration detention: its history, evolution, diffusion, legal status, consistency with democratic principles, and human impact. Dr Nethery's scholarship has been published in leading international journals. Her PhD thesis entitled *Immigration Detention in Australia* won the 2011 Isi Leibler Prize for the thesis that best advances knowledge on racism in Australia.

Adrian Kent is Professor of Quantum Physics at the Department of Applied Mathematics and Theoretical Physics, University of Cambridge, a Fellow of Wolfson College and Director of Studies in Mathematics at Darwin College. He is also a Distinguished Visiting Research Chair at the Perimeter Institute for Theoretical Physics, Fellow of the UK Institute of Physics, Founder Member of the Foundational Questions Institute and Charter Honorary Fellow of the John Bell Institute for the Foundations of Physics.

Adrian's research interests include the nature of physical reality, fundamental tests of quantum theory and its relationship to gravity, the properties of quantum information in space-time, and applications of fundamental physics to new forms of quantum and relativistic cryptography, communication and computation.

Heonik Kwon is a Senior Research Fellow in Social Anthropology at Trinity College, Cambridge. Author of prize-winning books on the sociocultural history of the Vietnam War, the Korean War and Asia's postcolonial Cold War more broadly, he previously taught in the London School of Economics and the University of Edinburgh. His new book, *Spirit Power* (2022), approaches Korea's Cold War experience from a religious historical angle. The present lecture draws upon his earlier co-authored book, *North Korea: Beyond Charismatic Power* (2012).

Jane Francis is a geologist by training, with research interests in understanding past climate change. She has undertaken research projects at the Universities of Southampton, London, Leeds and Adelaide, using fossil plants to determine the change from greenhouse to icehouse climates in the polar regions over the past 100 million years. She has undertaken over 15 scientific expeditions to the Arctic and Antarctica in search of fossil forests.

Jane is Director of the British Antarctic Survey, a research centre of the Natural Environment Research Council (UKRI-NERC). She is involved with international polar organisations, such as the Antarctic Treaty and European Polar Board, and on several advisory boards of national polar programmes.

Jane was appointed Dame Commander of the Order of St Michael and St George (DCMG) in recognition of services to UK polar science and diplomacy. She was also awarded the Polar Medal by H.M The Queen, and in 2018 became Chancellor of the University of Leeds.

Philip Jones is a physicist whose research centres on optical trapping and its use as a tool for probing a variety of nanoscopic, soft matter and biological materials. He is currently Professor of Physics at University College London where he leads the Optical Tweezers Group. In 2015 he co-authored the first textbook on optical tweezers *Optical Tweezers: Principles and Applications*, published by Cambridge University Press.

Arik Kershenbaum is a zoologist, Director of Studies, and Fellow at Girton College, University of Cambridge, and an expert on animal vocal communication, which he has researched for the past 14 years. He received his PhD at the University of Haifa in Israel, and holds a Higher Doctorate from the University of Cambridge. His first popular science book, *The Zoologist's Guide to the Galaxy* was a Times/Sunday Times Book of the Year, and received accolades from among others, Richard Dawkins and Lord Martin Rees, the Astronomer Royal. Dr Kershenbaum travels the world researching the nature of information in the communication of wolves, dolphins, primates, and other species, looking for indications of the similarities – and differences – with human speech. The evolution of different forms of animal communication is the subject of his second popular science book: *How Animals Talk*, which will be published in 2023.

Amrita Narlikar is the President of the German Institute for Global and Area Studies (GIGA). She is also a non-resident Senior Fellow at the Observer Research Foundation in Delhi, and non-resident Distinguished Fellow at the Australia-India Institute at the University of Melbourne. In 2021, she was elected to an Honorary Fellowship at Darwin College, University of Cambridge. Amrita is the author/editor of 13 volumes, the most recent of which is the co-edited How-Not-To Guide for International Relations, which was selected as a special issue (September 2022) to mark the centenary anniversary of *International Affairs*. Prior to moving to Germany, Amrita was Reader in International Political Economy at the University of Cambridge. She read for her M.Phil. and D.Phil. at Balliol College, Oxford, and held a Junior Research Fellowship at St John's College, Oxford. More information about her research and policy engagement, on https://www.giga-hamburg.de/en/the-giga/team/narlikar-amrita and on Twitter @AmritaNarlikar.

THE DARWIN COLLEGE LECTURES

The appeal of the annual Darwin College Lecture Series lies in its interdisciplinary content, with the humanities and sciences given an equal weighting. The lectures are delivered by obviously distinguished scholars who have the ability to communicate sometimes difficult concepts to a wide audience.

Now in its 38th year, the Lecture Series are available in book form and include all the topics listed here.

2015   *DEVELOPMENT*
       . . . things change, towards ever-increasing complexity
       eds Torsten Krude, Sara T Baker

2014   *PLAGUES*
       . . . infectious diseases to over-population and computer viruses
       eds Jonathan L Heeney, Sven Friedemann

2013   *FORESIGHT*
       . . . from: warfare, journalism, music, civilisations, space weather. . .
       eds Lawrence W Sherman, David Allan Feller

2012   *LIFE*
       . . . conventional and unconventional views on the meaning of life.
       eds William Brown, Andrew Fabian

2011   *BEAUTY*
       . . . entertaining, accessible and thought-provoking.
       eds Lauren Arrington, Zoë Leinhardt, Philip Dawid

2010   *RISK*
       . . . thinking about and trying to understand risk.
       eds Layla Skinns, Michael Scott, Tony Cox

2009   *DARWIN*
       thoughts, ideas, research and writings have impact. . .
       eds William Brown, Andrew Fabian

2008   *SERENDIPITY*
       . . .good fortune and the preparation of the mind.
       eds Mark de Rond, Iain Morley

2007   *IDENTITY*
       . . . what identity means across a number of academic disciplines.
       eds Giselle Walker, Elisabeth Leedham-Green

2006   *SURVIVAL*
       . . . strategies to assist existence.
       ed. Emily Shuckburgh

2005   *CONFLICT*
       . . . an inevitable part of the fabric of our existence?
       eds Martin Jones, Andrew Fabian

1993   *COLOUR: Art and Science*
        . . . a journey through the rewarding sensation of colour.
        eds Trevor Lamb, Janine Bourriau

1992   *WHAT IS INTELLIGENCE*
        . . . how we define intelligence?
        ed. Jean Khalfa

1991   *PREDICTING THE FUTURE*
        . . . humankind's obsessive urge to look beyond the present.
        eds Leo Howe, Alan Wain

1990   *UNDERSTANDING CATASTROPHE*
        . . . impact of catastrophic change on life on earth.
        ed. Janine Bourriau

1989   *WAYS OF COMMUNICATING*
        . . . this elusive aspect of our common humanity.
        eds David H. Mellor

1988   *DISCOVERIES*
        (unpublished)

1987   *THE FRAGILE ENVIRONMENT*
        . . . impact of the human species.
        eds L. E. Friday, Ronald Laskey

1986   *ORIGINS*
        . . . origins of the most fundamental features of our world.
        ed. Andrew Fabian

Index

Printed in Poland
by Amazon Fulfillment
Poland Sp. z o.o., Wrocław

13552227R00105